Pre-Calculus Workbook

3rd Edition

by Mary Jane Sterling

Pre-Calculus Workbook For Dummies®, 3rd Edition

Published by: **John Wiley & Sons, Inc.,** 111 River Street, Hoboken, NJ 07030-5774, www.wiley.com

Copyright © 2019 by John Wiley & Sons, Inc., Hoboken, New Jersey

Published simultaneously in Canada

No part of this publication may be reproduced, stored in a retrieval system or transmitted in any form or by any means, electronic, mechanical, photocopying, recording, scanning or otherwise, except as permitted under Sections 107 or 108 of the 1976 United States Copyright Act, without the prior written permission of the Publisher. Requests to the Publisher for permission should be addressed to the Permissions Department, John Wiley & Sons, Inc., 111 River Street, Hoboken, NJ 07030, (201) 748-6011, fax (201) 748-6008, or online at www.wiley.com/go/permissions.

Trademarks: Wiley, For Dummies, the Dummies Man logo, Dummies.com, Making Everything Easier, and related trade dress are trademarks or registered trademarks of John Wiley & Sons, Inc. and may not be used without written permission. All other trademarks are the property of their respective owners. John Wiley & Sons, Inc. is not associated with any product or vendor mentioned in this book.

LIMIT OF LIABILITY/DISCLAIMER OF WARRANTY: THE PUBLISHER AND THE AUTHOR MAKE NO REPRESENTATIONS OR WARRANTIES WITH RESPECT TO THE ACCURACY OR COMPLETENESS OF THE CONTENTS OF THIS WORK AND SPECIFICALLY DISCLAIM ALL WARRANTIES, INCLUDING WITHOUT LIMITATION WARRANTIES OF FITNESS FOR A PARTICULAR PURPOSE. NO WARRANTY MAY BE CREATED OR EXTENDED BY SALES OR PROMOTIONAL MATERIALS. THE ADVICE AND STRATEGIES CONTAINED HEREIN MAY NOT BE SUITABLE FOR EVERY SITUATION. THIS WORK IS SOLD WITH THE UNDERSTANDING THAT THE PUBLISHER IS NOT ENGAGED IN RENDERING LEGAL, ACCOUNTING, OR OTHER PROFESSIONAL SERVICES. IF PROFESSIONAL ASSISTANCE IS REQUIRED, THE SERVICES OF A COMPETENT PROFESSIONAL PERSON SHOULD BE SOUGHT. NEITHER THE PUBLISHER NOR THE AUTHOR SHALL BE LIABLE FOR DAMAGES ARISING HEREFROM. THE FACT THAT AN ORGANIZATION OR WEBSITE IS REFERRED TO IN THIS WORK AS A CITATION AND/OR A POTENTIAL SOURCE OF FURTHER INFORMATION DOES NOT MEAN THAT THE AUTHOR OR THE PUBLISHER ENDORSES THE INFORMATION THE ORGANIZATION OR WEBSITE MAY PROVIDE OR RECOMMENDATIONS IT MAY MAKE. FURTHER, READERS SHOULD BE AWARE THAT INTERNET WEBSITES LISTED IN THIS WORK MAY HAVE CHANGED OR DISAPPEARED BETWEEN WHEN THIS WORK WAS WRITTEN AND WHEN IT IS READ.

For general information on our other products and services, please contact our Customer Care Department within the U.S. at 877-762-2974, outside the U.S. at 317-572-3993, or fax 317-572-4002. For technical support, please visit https://hub.wiley.com/community/support/dummies.

Wiley publishes in a variety of print and electronic formats and by print-on-demand. Some material included with standard print versions of this book may not be included in e-books or in print-on-demand. If this book refers to media such as a CD or DVD that is not included in the version you purchased, you may download this material at http://booksupport.wiley.com. For more information about Wiley products, visit www.wiley.com.

Library of Congress Control Number: 2019931480

ISBN 978-1-119-50880-9 (pbk); ISBN 978-1-119-50881-6 (ebk); ISBN 978-1-119-50882-3 (ebk)

Manufactured in the United States of America

C10013326_082219

Contents at a Glance

Table of Contents

Introduction

I hope that you'll find this workbook to be a big help with pre-calculus. If you've gotten this far in your math career, congratulations! Many students choose to stop their math education after they complete Algebra II, but not you!

If you've picked up this book (and obviously you have, given that you're reading this sentence!), maybe some of the concepts in pre-calculus are giving you a hard time, or perhaps you just want more practice. Maybe you're deciding whether you even want to take pre-calculus at all. This book fits the bill for all those reasons. And it's here to encourage you on your pre-calculus adventure.

You'll find this workbook chock-full of valuable practice problems and explanations. In instances where you feel you may need a more thorough explanation, please refer to *Pre-Calculus For Dummies* by Mary Jane Sterling (Wiley). This book, however, is a great stand-alone workbook if you need extra practice or want to just brush up in certain areas.

About This Book

Don't let pre-calculus scare you. When you realize that you already know a whole bunch from Algebra I and Algebra II, you'll see that pre-calculus is really just using that old information in a new way. And even if you're scared, I'm here with you, so no need to panic. Before you get ready to start this new adventure, you need to know a few things about this book.

This book isn't a novel. It's not meant to be read in order from beginning to end. You can read any topic at any time, but it's structured it in such a way that it follows the "normal" curriculum. This is hard to do, because most states don't have state standards for what makes pre-calculus pre-calculus. Looking at a good sampling of curriculums, though, this should be a good representation of a pre-calculus course.

Here are two suggestions for using this book:

>> Look up what you need to know when you need to know it. The index and the table of contents direct you where to look.

>> Start at the beginning and read straight through. This way, you may be reminded of an old topic that you had forgotten (anything to get those math wheels churning inside your head). Besides, practice makes perfect, and the problems in this book are a great representation of the problems found in pre-calculus textbooks.

For consistency and ease of navigation, this book uses the following conventions:

>> Math terms are *italicized* when they're introduced or defined in the text.

>> Variables are *italicized* to set them apart from letters.

>> The symbol used when writing imaginary numbers is a lowercase *i*.

Foolish Assumptions

I don't assume that you love math the way I do, but I do assume that you picked this book up for a reason of your own. Maybe you want a preview of the course before you take it, or perhaps you need a refresher on the topics in the course, or maybe your kid is taking the course and you're trying to help him to be more successful.

Whatever your reason, I assume that you've encountered most of the topics in this book before, because for the most part, they review what you've seen in algebra or geometry.

Icons Used in This Book

Throughout this book you'll see icons in the margins to draw your attention to something important that you need to know.

EXAMPLE

You see this icon when I present an example problem whose solution I walk you through step by step. You get a problem and a detailed answer.

TIP

Tips are great, especially if you wait tables for a living! These tips are designed to make your life easier, which are the best tips of all!

TECHNICAL STUFF

The material following this icon is wonderful mathematics; it's closely related to the topic at hand, but it's not absolutely necessary for your understanding of the material being presented. You can take it or leave it — you'll be fine just taking note and leaving it behind as you proceed through the section.

REMEMBER

The Remember icon is used one way: It asks you to remember old material from a previous math course.

WARNING

Warnings are big red flags that draw your attention to common mistakes that may trip you up.

Beyond the Book

No matter how well you understand the concepts of algebra, you'll likely come across a few questions where you don't have a clue. Be sure to check out the free Cheat Sheet for a handy guide that covers tips and tricks for answering pre-calculus questions. To get this Cheat Sheet, simply go to www.dummies.com and enter "Pre-Calculus Workbook For Dummies" in the Search box.

The online practice that comes free with this book contains over 300 questions so you can really hone your pre-calculus skills! To gain access to the online practice, all you have to do is register. Just follow these simple steps:

1. **Find your PIN access code located on the inside front cover of this book.**

2. **Go to Dummies.com and click** Activate Now.

3. **Find your product (***Pre-Calculus Workbook For Dummies***, 3rd Edition with Online Practice) and then follow the on-screen prompts to activate your PIN.**

Now you're ready to go! You can go back to the program at testbanks.wiley.com as often as you want — simply log on with the username and password you created during your initial login. No need to enter the access code a second time.

Tip: If you have trouble with your PIN or can't find it, contact Wiley Product Technical Support at 877-762-2974 or go to support.wiley.com.

Where to Go from Here

Pick a starting point in the book and go practice the problems there. If you'd like to review the basics first, start at Chapter 1. If you feel comfy enough with your algebra skills, you may want to skip that chapter and head over to Chapter 2. Most of the topics there are reviews of Algebra II material, but don't skip over something because you think you have it under control. You'll find in pre-calculus that the level of difficulty in some of these topics gets turned up a notch or two. Go ahead — dive in and enjoy the world of pre-calculus!

If you're ready for another area of mathematics, look for a couple more of my titles: *Trigonometry For Dummies* and *Linear Algebra For Dummies*.

1

Setting the Foundation: The Nuts and Bolts of Pre-Calculus

Pre-calculus is really just another stop on the road to calculus. The chapters in this part begin with a review of the basics: using the order of operations, solving and graphing equations and inequalities, and using the distance and midpoint formulas. Some new material pops up in the form of interval notation, so be sure and check that out. As you move on to real numbers you find yourself focusing on radicals. Everything you ever wanted to know about functions is covered here: graphing and transforming parent graphs, dealing with rational functions, and piecewise functions. You'll see how to perform operations on functions and how to find the inverse. Then you move on to solving higher-degree polynomials using techniques like factoring, completing the square, and the quadratic formula. You also find out how to graph these complicated polynomials. Lastly, you discover exponential and logarithmic functions and what you're expected to know about them.

Chapter **1**

Preparing for Pre-Calculus

P re-calculus is the stepping stone for calculus. It's the final stepping stone after all those years of math: algebra I, geometry, algebra II, and trigonometry. Now all you need is pre-calculus to get to that ultimate goal — calculus. And as you may recall from your algebra II class, you were subjected to much of the same material you saw in algebra and even pre-algebra (just a couple steps up in terms of complexity — but really the same stuff). Pre-calculus begins with certain concepts that you need to be successful in any mathematics course.

If you feel you're already an expert at everything algebra, feel free to skip past this chapter and get the full swing of pre-calculus going. If you have any doubts or concerns, however, you may want to review; read on.

TIP

If you don't remember some of the concepts discussed in this chapter, or even in this book, you can pick up another *For Dummies* math book for review. The fundamentals are important. That's why they're called fundamentals. Take the time now to review and save yourself frustration and possible math errors in the future!

Reviewing Order of Operations: The Fun in Fundamentals

You can't put on your sock after you put on your shoe, can you? At least, you shouldn't! The same concept applies to mathematical operations. There's a specific order as to which operation you perform first, second, third, and so on. At this point, it should be second nature, but because the concept is so important (especially when you start doing more complex calculations), a quick review is worth it, starting with everyone's favorite mnemonic device.

TIP

Please excuse who? Oh, yeah, you remember this one — my dear Aunt Sally! The old mnemonic still stands, even as you get into more complicated problems. Please Excuse My Dear Aunt Sally is a mnemonic for the acronym PEMDAS, which stands for

» Parentheses (including absolute value, brackets, fraction lines, and radicals)

» Exponents (and roots)

» Multiplication and Division (from left to right)

» Addition and Subtraction (from left to right)

The order in which you solve algebraic problems is very important. Always compute what's in the parentheses first, then move on to the exponents, followed by the multiplication and division (from left to right), and finally, the addition and subtraction (from left to right).

TECHNICAL STUFF

You should also have a good grasp on the properties of equality. If you do, you'll have an easier time simplifying expressions. Here are the properties:

» **Reflexive property:** $a = a$. For example, $4 = 4$.

» **Symmetric property:** If $a = b$, then $b = a$. For example, if $2 + 8 = 10$, then $10 = 2 + 8$.

» **Transitive property:** If $a = b$ and $b = c$, then $a = c$. For example, if $2 + 8 = 10$ and $10 = 5 \cdot 2$, then $2 + 8 = 5 \cdot 2$.

» **Commutative property of addition:** $a + b = b + a$. For example, $3 + 4 = 4 + 3$.

» **Commutative property of multiplication:** $a \cdot b = b \cdot a$. For example, $3 \cdot 4 = 4 \cdot 3$.

» **Associative property of addition:** $a + (b + c) = (a + b) + c$. For example, $3 + (4 + 5) = (3 + 4) + 5$.

» **Associative property of multiplication:** $a \cdot (b \cdot c) = (a \cdot b) \cdot c$. For example, $3 \cdot (4 \cdot 5) = (3 \cdot 4) \cdot 5$.

» **Additive identity:** $a + 0 = a$. For example, $4 + 0 = 4$.

» **Multiplicative identity:** $a \cdot 1 = a$. For example, $-18 \cdot 1 = -18$.

» **Additive inverse property:** $a + (-a) = 0$. For example, $5 + (-5) = 0$.

» **Multiplicative inverse property:** $a \cdot \dfrac{1}{a} = 1$, as long as $a \neq 0$. For example, $-2 \cdot \left(-\dfrac{1}{2}\right) = 1$.

» **Distributive property:** $a(b+c) = a \cdot b + a \cdot c$. For example, $5(4+3) = 5 \cdot 4 + 5 \cdot 3$.

» **Multiplicative property of zero:** $a \cdot 0 = 0$. For example, $4 \cdot 0 = 0$.

» **Zero product property: If** $a \cdot b = 0$, **then** $a = 0$ or $b = 0$. For example, if $x(2x - 3) = 0$, then $x = 0$ or $(2x - 3) = 0$.

Following are a couple examples so you can see the order of operations and the properties of equality in action before diving into some practice questions.

EXAMPLE

Q. Simplify: $\dfrac{6^2 - 4\left(3 - \sqrt{20+5}\right)^2}{|4-8|}$

A. The answer is 5.

Following the order of operations, simplify everything in parentheses first. (Remember that radicals and absolute value bars act like parentheses, so do operations within them first before simplifying the radicals or taking the absolute value.)

$$= \frac{6^2 - 4\left(3 - \sqrt{25}\right)^2}{|-4|}$$

Simplify the parentheses by taking the square root of 25 and subtracting the result from 3; find the absolute value of −4:

$$= \frac{6^2 - 4(3-5)^2}{|-4|} = \frac{6^2 - 4(-2)^2}{4}$$

Now you can deal with the exponents by squaring the 6 and the −2:

$$= \frac{36 - 4 \cdot 4}{4}$$

Although they're not written, parentheses are implied around the terms above and below a fraction bar. Therefore, you must simplify the numerator and denominator before dividing the terms following the order of operations:

$$= \frac{36 - 16}{4} = \frac{20}{4} = 5$$

Q. Simplify: $\dfrac{\left(\dfrac{1}{8} + \dfrac{1}{3}\right) + \dfrac{3}{8}}{\dfrac{1}{6} + \dfrac{1}{9}}$

A. The answer is 3.

Using the associative property and the commutative property of addition, rewrite the expression to make the fractions easier to add. Then add the fractions with the common denominators.

$$= \frac{\left(\dfrac{1}{8} + \dfrac{3}{8}\right) + \dfrac{1}{3}}{\dfrac{1}{6} + \dfrac{1}{9}} = \frac{\left(\dfrac{4}{8}\right) + \dfrac{1}{3}}{\dfrac{1}{6} + \dfrac{1}{9}}$$

Then reduce the resulting fraction and change the fractions in the numerator and denominator to equivalent fractions with common denominators:

$$= \frac{\dfrac{1}{2} + \dfrac{1}{3}}{\dfrac{1}{6} + \dfrac{1}{9}} = \frac{\dfrac{3}{6} + \dfrac{2}{6}}{\dfrac{3}{18} + \dfrac{2}{18}}$$

Adding the fractions, you get:

$$= \frac{\dfrac{5}{6}}{\dfrac{5}{18}}$$

To simplify the complex fraction, you multiply the numerator by the reciprocal of the denominator:

$$= \frac{\cancel{5}}{\cancel{6}} \cdot \frac{\cancel{18}^{3}}{\cancel{5}} = \frac{3}{1} = 3$$

1 Simplify: $\dfrac{3\sqrt{(4-6)^2+[2-(-1)]^2}}{|-3-(-1)|}$

2 Simplify: $\dfrac{|-3|-|2|+(-1)}{|-7+2|}$

3 Simplify: $\left(2^3-3^2\right)^4(-5)$

4 Simplify: $\dfrac{|5(1-4)+6|}{3\left(-\dfrac{1}{6}+\dfrac{1}{3}\right)-\dfrac{1}{2}}$

Keeping Your Balance While Solving Equalities

Just as simplifying expressions is a basic process in pre-algebra, solving for variables is the basis of algebra. And both are essential to the more complex concepts covered in pre-calculus.

Solving linear equations with the general format of $ax+b=c$, where a, b, and c are constants, is relatively easy using the properties of numbers. The goal, of course, is to isolate the variable, x.

REMEMBER

One type of equation you can't forget is the absolute value equation. The *absolute value* of a number is defined as its distance from 0. In other words, $|x|=\begin{cases}x, x\geq 0\\-x, x<0\end{cases}$. This definition is a piecewise function with two rules: one where the quantity inside the absolute value bars is positive and another where it's negative. To solve these equations, you must isolate the absolute value term and then set the quantity inside the absolute value bars to the positive and negative values (see the second example question that follows).

Check out the following examples or skip ahead to the practice questions if you think you're ready to tackle them.

EXAMPLE

Q. Solve for x: $3(2x-4) = x-2(3-2x)$

A. $x = 6$

First, using the distributive property, distribute the 3 and the −2 to get $6x-12 = x-6+4x$. Then combine like terms and solve using algebra, like so: $6x-12 = 5x-6$ giving you $x-12 = -6$, and, finally, $x = 6$.

Q. Solve for x: $|x-3|+(-16) = -12$

A. $x = 7$ or -1

Isolate the absolute value by adding 16 to each side, giving you $|x-3| = 4$. One solution comes when you assume that the quantity inside the absolute value bars is positive: $x-3 = 4$. This gives you the answer $x = 7$. The second solution comes from assuming that the quantity inside the absolute value bars is negative: $-(x-3) = 4$. This becomes $-x+3 = 4$, then $-x = 1$, and finally $x = -1$.

---

5 Solve: $3-6\left[2-4x(x+3)\right] = 3x(8x+12)+27$

6 Solve: $\dfrac{x}{2}+\dfrac{x-2}{4} = \dfrac{x+4}{2}$

---

7 Solve: $|x-3|+|3x+2| = 4$

8 Solve: $3-4(2-3x) = 2(6x+2)$

---

9 Solve: $2|x-3|+12=6$

10 Solve: $3(2x+5)+10=2(x+10)+4x+5$

When Your Image Really Counts: Graphing Equalities and Inequalities

Graphs are visual representations of mathematical equations. In pre-calculus, you'll be introduced to many new mathematical equations and then be expected to graph them. You will have plenty of practice graphing these equations when you read the material involving the more complex equations. In the meantime, it's important to practice the basics: graphing linear equations and inequalities.

TECHNICAL STUFF

The graphs of linear equations and inequalities exist on the *Cartesian coordinate system,* which is made up of two axes: the horizontal, or *x*-axis, and the vertical, or *y*-axis. Each point on the coordinate plane is called a *Cartesian coordinate pair* and has an *x* coordinate and a *y* coordinate. The notation for any point on the coordinate plane looks like this: (x, y). A set of these ordered pairs that can be graphed on a coordinate plane is called a *relation.* The *x* values of a relation are its *domain,* and the *y* values are its *range.* For example, the domain of the relation $R = \{(2, 4), (-5, 3), (1, -2)\}$ is $\{2, -5, 1\}$, and the range is $\{4, 3, -2\}$.

You can graph a linear equation using two points or by using the slope-intercept form. The same can be used when graphing linear inequalities. These approaches are reviewed in the following sections.

Graphing with two points

To graph a line using two points, choose two numbers and plug them into the equation to solve for the range (*y*) values. After you plot these points (x, y) on the coordinate plane, you can draw the line through the points.

A nice alternative is to use the two intercepts, the points that fall on the x- or y-axes. To find the x-intercept (x, 0), plug in 0 for y and solve for x. To find the y-intercept (0, y), plug in 0 for x and solve for y. For example, to find the intercepts of the linear equation $2x + 3y = 12$, start by plugging in 0 for y: $2x + 3(0) = 12$. Then, using properties of numbers, solve for x and you get $x = 6$. So the x-intercept is (6, 0). For the y-intercept, plug in 0 for x and solve for y: $2(0) + 3y = 12$ which give you $y = 4$. Therefore, the y-intercept is (0, 4). At this point, you can plot those two points and connect them to graph the line, because, as you learned in geometry, two points make a line. See the resulting graph in Figure 1-1.

Graphing by using the slope-intercept form

The slope-intercept form of a linear equation gives a great deal of helpful information in a nice package. The equation $y = mx + b$ immediately gives you the y-intercept (b); it also gives you the slope (m). *Slope* is a fraction that gives you the rise over the run. To change equations that aren't written in slope-intercept form, you simply solve for y. For example, if you use the linear equation $2x + 3y = 12$, you start by subtracting 2x from each side: $3y = -2x + 12$. Next, you divide all the terms by 3 giving you $y = -\frac{2}{3}x + 4$. Now that the equation is in slope-intercept form, you know that the y-intercept is 4, and you can plot this point on the coordinate plane. Then you can use the slope to plot a second point. From the slope-intercept equation, you know that the slope is $-\frac{2}{3}$. This tells you that the rise is –2 and the run is 3. From the point (0, 4), plot the point 2 down and 3 to the right. In other words, (3, 2). Lastly, connect the two points to graph the line. The resulting graph is identical to Figure 1-1.

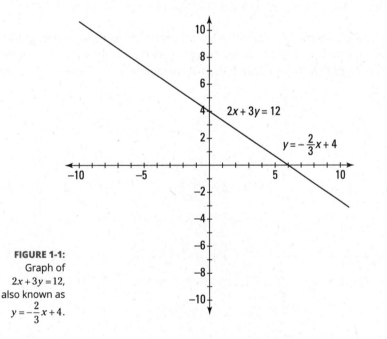

FIGURE 1-1:
Graph of $2x + 3y = 12$, also known as $y = -\frac{2}{3}x + 4$.

Graphing inequalities

REMEMBER

Similar to graphing linear equations, graphing linear inequalities begins with plotting two points. However, because *inequalities* are used for comparisons — greater than, less than, or equal to — you have two more questions to answer after finding two points:

» Is the line dashed (< or >) or solid (≤ or ≥)?

» Do you shade under the line ($y <$ or $y \leq$) or above the line ($y >$ or $y \geq$)?

Here's an example of graphing an inequality followed by a few practice questions.

EXAMPLE

Q. Sketch the graph of the inequality: $3x - 2y > 4$

A. Put the inequality into slope-intercept form by subtracting $3x$ from each side of the equation to get $-2y > -3x + 4$ and then dividing each term by -2 to get $y < \frac{3}{2}x - 2$. (*Remember:* When you multiply or divide an inequality by a negative, you need to reverse the inequality.) From the resulting statement, you can find the y-intercept, -2, and the slope, $\frac{3}{2}$. Use this information to graph two points by using the slope-intercept form. Next, decide the nature of the line (solid or dashed). Because the inequality is strict, the line is dashed. Graph the dashed line so you can decide where to shade. Because $y < \frac{3}{2}x - 2$ is a less-than inequality, shade below the dashed line, as shown in the following figure.

This method works only if the boundary line is first converted to slope–intercept form. An alternative is to graph the boundary line using any method and then use a sample point, such as (0, 0), to determine which half-plane to shade.

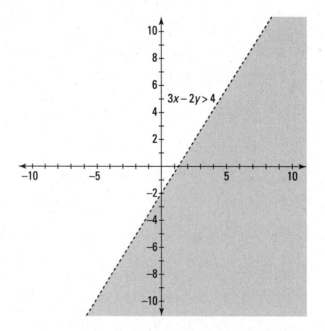

11 Sketch the graph of $\frac{4}{3}(6x+2y)=16$.

12 Sketch the graph of $\frac{5x+4y}{2}\geq 6$.

13 Sketch the graph of $4x+5y\geq 2(3y+2x+4)\cdot$

14 Sketch the graph of $x-3y=4-2y-y$.

Using Graphs to Find Distance, Midpoint, and Slope

Graphs are more than just pretty pictures. From a graph, it's possible to choose two points and then figure out the distance between them, the midpoint of the segment connecting them, and the slope of the line running through them. As graphs become more complex in both pre-calculus and calculus, you're asked to find and use all three of these pieces of information. Aren't you lucky?

Finding the distance

TECHNICAL STUFF

Distance refers to how far apart two things are. In this case, you're finding the distance between two points. Knowing how to calculate distance is helpful for when you get to conics (see Chapter 12). To find the distance between two points (x_1, y_1) and (x_2, y_2), use the following formula: $d=\sqrt{(x_1-x_2)^2+(y_1-y_2)^2}$.

Calculating the midpoint

TECHNICAL STUFF

The *midpoint* is the middle of a segment. This concept also comes up in conics (see Chapter 12) and is ever so useful for all sorts of other pre-calculus calculations. To find the coordinates of the midpoint, M, of the points (x_1, y_1) and (x_2, y_2), you just need to average the x and y values and express them as an ordered pair, like so: $M = \left(\dfrac{x_1 + x_2}{2}, \dfrac{y_1 + y_2}{2} \right)$.

Discovering the slope

TECHNICAL STUFF

Slope is a key concept for linear equations, but it also has applications for trigonometric functions and is essential for differential calculus. *Slope* describes the steepness of a line on the coordinate plane (think of a ski slope). Use this formula to find the slope, m, of the line (or segment) connecting the two points (x_1, y_1) and (x_2, y_2): $m = \dfrac{y_2 - y_1}{x_2 - x_1}$.

Note: Positive slopes move upward as you move from left to right. Negative slopes move downward as you move from left to right. Horizontal lines have a slope of 0, and vertical lines have an undefined slope.

Following is an example question for your reviewing pleasure. Look it over and then try your hand at the practice questions.

Q. Find the distance, slope, and midpoint of $\overline{AB}$.

EXAMPLE

A. The distance is $\sqrt{65}$, the slope is $\dfrac{4}{7}$, and the midpoint is $\left(\dfrac{3}{2}, 1 \right)$.

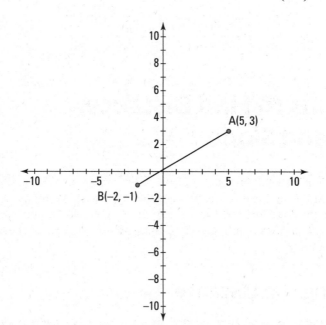

Plug the *x* and *y* values into the distance formula and, following the order of operations, simplify the terms under the radical (keeping in mind the implied parentheses of the radical itself).

$$d = \sqrt{\left(5-(-2)\right)^2 + \left(3-(-1)\right)^2}$$
$$= \sqrt{7^2 + 4^2} = \sqrt{49+16} = \sqrt{65}$$

Because 65 doesn't contain any perfect squares as factors, this is as simple as you can get.

To find the midpoint, plug the points into the midpoint formula, and simplify using the order of operations.

$$M = \left(\frac{5+(-2)}{2}, \frac{3+(-1)}{2}\right)$$
$$= \left(\frac{3}{2}, \frac{2}{2}\right) = \left(\frac{3}{2}, 1\right)$$

To find the slope, use the formula, plug in your *x* and *y* values, and use the order of operations to simplify.

$$m = \frac{-1-3}{-2-5}$$
$$= \frac{-4}{-7} = \frac{4}{7}$$

 Find the length of segment *CD*, where *C* is (−2, 4) and *D* is (3, −1).

 Find the midpoint of segment *EF*, where *E* is (3, −5) and *F* is (7, 5).

17 Find the slope of line *GH*, where *G* is $(-3, -5)$ and *H* is $(-3, 4)$.

18 Find the perimeter of triangle *CAT*.

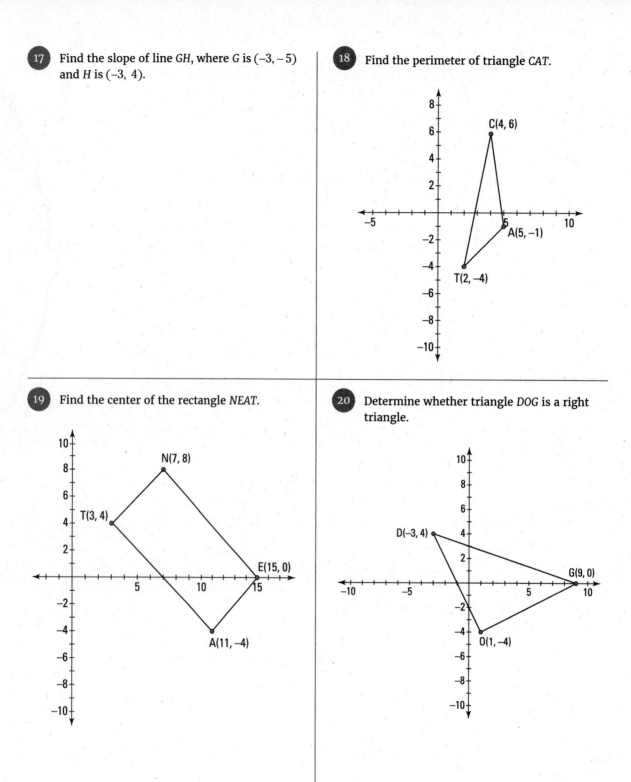

19 Find the center of the rectangle *NEAT*.

20 Determine whether triangle *DOG* is a right triangle.

Answers to Problems on Fundamentals

Following are the answers to questions dealing with pre-calculus fundamentals.

(1) Simplify $\dfrac{3\sqrt{(4-6)^2+[2-(-1)]^2}}{|-3-(-1)|}$. The answer is $\dfrac{3\sqrt{13}}{2}$.

Start by simplifying everything in the parentheses. Next, simplify the exponents. Finally, add the remaining terms. Here's what your math should look like:

$$\frac{3\sqrt{(4-6)^2+[2-(-1)]^2}}{|-3-(-1)|}=\frac{3\sqrt{(-2)^2+[3]^2}}{|-3+1|}=\frac{3\sqrt{4+9}}{|-2|}=\frac{3\sqrt{13}}{2}$$

(2) Simplify $\dfrac{|-3|-|2|+(-1)}{|-7+2|}$. The answer is 0.

Recognizing that the absolute value in the denominator acts as parentheses, add the -7 and 2 inside there first. Then rewrite the absolute value of each. Next, add the terms in the numerator. Finally, recognize that $\dfrac{0}{5}$ equals 0.

$$\frac{|-3|-|2|+(-1)}{|-7+2|}=\frac{|-3|-|2|+(-1)}{|-5|}=\frac{3-2-1}{5}=\frac{0}{5}=0$$

(3) Simplify $\left(2^3-3^2\right)^4(-5)$. The answer is -5.

Begin by computing the exponential terms in the parentheses. Next, simplify the parentheses by subtracting. Find the power of the result and multiply the result, 1, by -5.

$$\left(2^3-3^2\right)^4(-5)=(8-9)^4(-5)=(-1)^4(-5)=1(-5)=-5$$

(4) Simplify $\dfrac{|5(1-4)+6|}{3\left(-\dfrac{1}{6}+\dfrac{1}{3}\right)-\dfrac{1}{2}}$. The answer is undefined.

Start by simplifying the parentheses in the numerator and denominator. Next, multiply the terms in the numerator and denominator. Then add the terms in the absolute value bars in the numerator and subtract the terms in the denominator. Take the absolute value of -9 to simplify the numerator. Finally, realizing that you can't have 0 in the denominator, you see that the resulting fraction is undefined.

$$\frac{|5(1-4)+6|}{3\left(-\dfrac{1}{6}+\dfrac{1}{3}\right)-\dfrac{1}{2}}=\frac{|5(-3)+6|}{3\left(\dfrac{1}{6}\right)-\dfrac{1}{2}}=\frac{|-15+6|}{\dfrac{1}{2}-\dfrac{1}{2}}=\frac{|-9|}{0}=\frac{9}{0},\text{ which is undefined.}$$

(5) Solve $3-6[2-4x(x+3)]=3x(8x+12)+27$. The answer is $x=1$.

Lots of parentheses in this one! Get rid of them by distributing terms. Start by distributing the $-4x$ on the left side over $(x+3)$ and the $3x$ on the right side over $(8x+12)$. Next, distribute the -6 over the remaining parentheses on the left side of the equation. Combine like terms on the left side. To isolate x onto one side, subtract $24x^2$ from both sides. Subtracting $36x$ from each side and adding 9 to both sides results in an equation where you can divide each side by 36, giving you the solution: $x = 1$.

$$3 - 6\left[2 - 4x(x+3)\right] = 3x(8x+12) + 27$$
$$3 - 6\left[2 - 4x^2 - 12x\right] = 24x^2 + 36x + 27$$
$$3 - 12 + 24x^2 + 72x = 24x^2 + 36x + 27$$
$$24x^2 + 72x - 9 = 24x^2 + 36x + 27$$
$$72x - 9 = 36x + 27$$
$$36x = 36$$
$$x = 1$$

6 Solve $\dfrac{x}{2} + \dfrac{x-2}{4} = \dfrac{x+4}{2}$. The answer is $x = 10$.

Multiply through by the common denominator, 4, to eliminate the fractions altogether. Then solve like normal by combining like terms and isolating x. Here's the math:

$$\frac{x}{2} + \frac{x-2}{4} = \frac{x+4}{2}, \ 4\left[\frac{x}{2} + \frac{x-2}{4} = \frac{x+4}{2}\right], \ 2x + x - 2 = 2x + 8, \ 3x - 2 = 2x + 8, \ x = 10$$

7 Solve $|x-3| + |3x+2| = 4$. The answer is $x = -\dfrac{3}{4}, -\dfrac{1}{2}$.

So you have two absolute value terms? Remember that absolute value means the distance from 0, so you have to consider all the possibilities to solve this problem. In other words, you have to consider and try four different possibilities: both absolute values are positive, both are negative, the first is positive and the second is negative, and the first is negative and the second is positive.

TECHNICAL STUFF

When you have multiple absolute value terms in a problem, not all the possibilities will work. As you calculate the possibilities, you may create what math people call *extraneous solutions*. These are actually false solutions that don't work in the original equation. You create extraneous solutions when you change the format of an equation. To be sure a solution is real and not extraneous, you need to plug your answer into the original equation to check it. Time to try the possibilities:

- **Positive/positive:** $(x-3) + (3x+2) = 4$, $4x - 1 = 4$, $4x = 5$, $x = \dfrac{5}{4}$. Plugging this answer back into the original equation, you get $\left|\dfrac{5}{4} - 3\right| + \left|3\left(\dfrac{5}{4}\right) + 2\right| = 4$. This simplifies to $\dfrac{7}{4} + \dfrac{23}{4} = \dfrac{30}{4} = \dfrac{15}{2} = 4$. Nope! You have an extraneous solution.

- **Negative/negative:** $-(x-3) - (3x+2) = 4$, $4x + 1 = 4$, $4x = 3$, $x = -\dfrac{3}{4}$. Plug that into the original equation and you get $4 = 4$. Voilà! Your first solution.

- **Positive/negative:** $(x-3) - (3x+2) = 4$, $-2x - 5 = 4$, $-2x = 9$, $x = -\dfrac{9}{2}$. Put it back into the original equation, and you get $19 = 4$. Nope, again — that's another extraneous solution.

- **Negative/positive:** $-(x-3) + (3x+2) = 4$, $2x + 5 = 4$, $2x = -1$, $x = -\dfrac{1}{2}$. Into the original equation it goes, and you get $4 = 4$. Ta da! Your second solution.

8 Solve $3 - 4(2 - 3x) = 2(6x + 2)$. The answer is no solution.

Distribute over the parentheses on each side: $3 - 8 + 12x = 12x + 4$. Combine like terms to get $-5 + 12x = 12x + 4$. When you subtract $12x$ from each side you get $-5 = 4$, which is false. Consequently, there is no solution.

9 Solve $2|x - 3| + 12 = 6$. The answer is no solution.

Isolate the absolute value by subtracting 12 from each side and then dividing by 2: $|x - 3| = -3$. Because an absolute value must be positive or zero, there's no solution to satisfy this equation.

10 Solve $3(2x + 5) + 10 = 2(x + 10) + 4x + 5$. The answer is all real numbers.

Begin by distributing over the parentheses on each side: $6x + 15 + 10 = 2x + 20 + 4x + 5$. Next, combine like terms on each side: $6x + 25 = 6x + 25$. Subtracting $6x$ from each side gives you $25 = 25$. This is a true statement. All real numbers would satisfy this equation.

11 Sketch the graph of $\frac{4}{3}(6x + 2y) = 16$. See the graph for the answer.

Using slope-intercept form, multiply both sides of the equation by the reciprocal of $\frac{4}{3}$, which is $\frac{3}{4}$: $\frac{3}{4} \cdot \frac{4}{3}(6x + 2y) = \frac{3}{4} \cdot 16$. Doing so gives you $6x + 2y = 12$. Next, solve for y by subtracting $6x$ from each side and dividing by 2, resulting in $y = -3x + 6$. Now, because $y = -3x + 6$ is in slope-intercept form, you can identify the slope (-3) and y-intercept (6). Use these to graph the equation. Start at the y-intercept (0, 6) and move to the right 1 unit and then down 3 units to find another point. Draw a line through the two points.

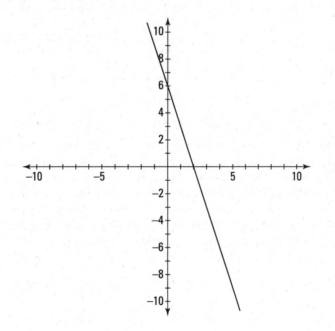

(12) Sketch the graph of $\dfrac{5x+4y}{2} \geq 6$. See the graph for the answer.

Start by multiplying both sides of the equation by 2, giving you $5x + 4y \geq 12$. Next, isolate y by subtracting $5x$ from each side and dividing by 4: $4y \geq -5x + 12$ and then $y \geq -\dfrac{5}{4}x + 3$. Now that you have slope-intercept form, you can graph the corresponding line $y = -\dfrac{5}{4}x + 3$. And because $y \geq -\dfrac{5}{4}x + 3$ is a greater-than-or-equal-to inequality, draw a solid line and shade above the line.

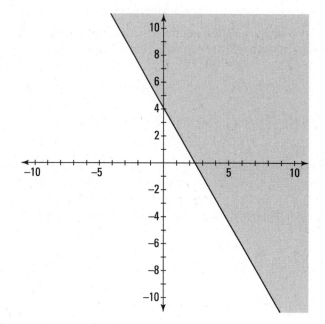

(13) Sketch the graph of $4x + 5y \geq 2(3y + 2x + 4)$. See the graph for the answer.

First things first: Get the equation into slope-intercept form by distributing the 2 on the left side. Next, isolate y by subtracting $4x$ from each side, subtracting y from each side, and then dividing by -1 (don't forget to switch your inequality sign, though). Here's what your work should look like: $4x + 5y \geq 2(3y + 2x + 4)$; $4x + 5y \geq 6y + 4x + 8$; $5y \geq 6y + 8$; $-y \geq 8$; $y \leq -8$. Because there's no x term, you can think of the slope-intercept form as containing $0x$, which tells you that the slope is 0. Therefore, the resulting line is a horizontal line at -8. Because the inequality is less than or equal to, you shade below a solid line.

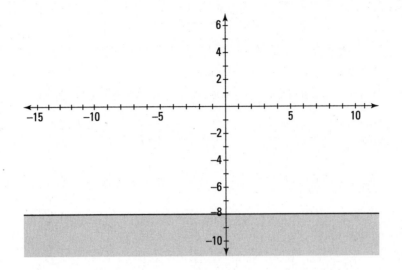

(14) Sketch the graph of $x - 3y = 4 - 2y - y$. See the graph for the answer.

Simplify the equation to put it in slope–intercept form. Combine like terms and add $3y$ to each side: $x - 3y = 4 - 3y$, $x = 4$. Here, the resulting line is a vertical line at 4.

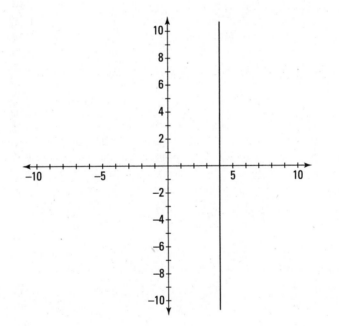

(15) Find the length of segment CD, where C is $(-2, 4)$ and D is $(3, -1)$. The answer is $d = 5\sqrt{2}$.

Using the distance formula, plug in the x and y values: $d = \sqrt{(-2-3)^2 + (4-(-1))^2}$. Then, simplify using the order of operations: $d = \sqrt{(-5)^2 + (5)^2} = \sqrt{25+25} = \sqrt{50} = 5\sqrt{2}$.

(16) Find the midpoint of segment EF, where E is $(3, -5)$ and F is $(7, 5)$. The answer is $M = (5, 0)$.

Using the midpoint formula, you get $M = \left(\dfrac{3+7}{2}, \dfrac{-5+5}{2}\right)$. Simplify from there to find that $M = \left(\dfrac{10}{2}, \dfrac{0}{2}\right) = (5, 0)$.

(17) Find the slope of line GH, where G is $(-3, -5)$ and H is $(-3, 4)$. The answer is the slope is undefined.

Using the formula for slope, plug in the x and y values for the two points: $m = \dfrac{4-(-5)}{-3-(-3)} = \dfrac{9}{0}$. This equation simplifies to $-9/0$, which is undefined.

(18) Find the perimeter of triangle CAT. The answer is $8\sqrt{2} + 2\sqrt{26}$.

To find the perimeter, you need to calculate the distance on each side of the triangle, which means you have to find the lengths of CA, AT, and TC. Plugging the values of x and y for each point into the distance formula, you find that the distances are as follows: $CA = 5\sqrt{2}$, $AT = 3\sqrt{2}$, and $TC = 2\sqrt{26}$. Adding like terms gives you the perimeter of $8\sqrt{2} + 2\sqrt{26}$.

(19) Find the center of the rectangle $NEAT$. The answer is $(9, 2)$.

The trick to this one is to realize that if you can find the midpoint of one of the rectangle's diagonals, you can identify the center of the rectangle. Easy, huh? So, by using the diagonal NA, you can find the midpoint and thus the center: $M = \left(\dfrac{7+11}{2}, \dfrac{8+(-4)}{2}\right) = \left(\dfrac{18}{2}, \dfrac{4}{2}\right) = (9, 2)$.

(20) Determine whether triangle DOG is a right triangle. The answer is yes.

This problem forces you to recall that right triangles have one set of perpendicular sides (which form that right angle) and that perpendicular sides have negative reciprocal slopes. In other words, if you multiply their slopes together, you get -1. With that information in your head, all you have to do is find the slopes of the lines that appear to be perpendicular. If they multiply to equal -1, then you know you have a right triangle.

Start by finding the slope of DO: $m = \dfrac{4-(-4)}{-3-1} = \dfrac{8}{-4} = -2$. Next, find the slope of OG: $m = \dfrac{-4-0}{1-9} = \dfrac{-4}{-8} = \dfrac{1}{2}$.

By multiplying the two slopes together, you find that they equal -1, indicating that you have perpendicular lines: $-2 \cdot \dfrac{1}{2} = -1$. Therefore, triangle DOG is a right triangle.

IN THIS CHAPTER

» **Finding solutions to equations
  with inequalities**

» **Using interval notation to express
  inequality**

» **Simplifying radicals and exponents**

» **Rationalizing the denominator**

Chapter **2**

Real Numbers Come Clean

I f fundamentals such as the order of operations (see Chapter 1) are the foundation of your pre-calculus house, then the skills you pick up in Algebra I and II are the mortar between your pre-calculus bricks. You need this knowledge if you're going to move forward in your mathematics studies. Without it, your pre-calculus house is going to take a beating. Never fear, though. This chapter is here to help refresh your memory.

In this chapter, I assume that you are well prepared in your basic algebra skills, so I review only the more challenging algebra concepts. In addition to reviewing inequalities, radicals, and exponents, I also introduce interval notation. (*Note:* If you feel confident with the other review sections in this chapter, feel free to skip ahead in the book, but make sure you practice some of the interval notation problems before moving on to Chapter 3.)

Solving Inequalities

Inequalities are mathematical sentences that indicate that two expressions are something other than just equal. They're expressed using the following symbols:

Greater than: >

Greater than or equal to: ≥

Less than: <

Less than or equal to: ≤

Solving equations with inequalities is almost exactly the same as solving equations with equalities. There's just one key exception: multiplying and dividing by negative numbers.

REMEMBER

When you multiply or divide each side of an inequality by a negative number, you must switch the direction of the inequality symbol. In other words, < becomes > and vice versa.

This is also a good time to put together two key concepts: inequalities and absolute values, or *absolute value inequalities*. With these, you need to remember that absolute values have two possible solutions: one when the quantity in the absolute value bars is positive, and one when it's negative. Therefore, you have to solve for these two possible solutions.

TECHNICAL STUFF

The easiest way to solve for the two possible solutions in linear absolute value inequalities is to drop the absolute value bars and apply these rules:

$|ax \pm b| < c$ becomes $ax \pm b < c$ AND $ax \pm b > -c$ which can also be written $-c < ax \pm b < c$

$|ax \pm b| > c$ becomes $ax \pm b > c$ OR $ax \pm b < -c$

These rules also apply to less-than-or-equal-to and greater-than-or-equal-to.

The solutions for these absolute value inequalities can be expressed graphically, as you can see in Figure 2-1. Note that the numbers representing the empty dots aren't included in the solutions.

FIGURE 2-1:
Graphical
solutions for
$|ax \pm b| < c$ and
$|ax \pm b| > c$.

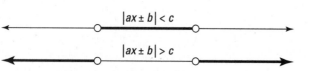

WARNING

Watch out for absolute value inequalities involving negative numbers. Here are two possible scenarios you may encounter:

>> **If the absolute value is less than or equal to a negative number, a solution doesn't exist.** Because an absolute value must be positive or zero, it can never be less than a negative number. For example, $|2x + 3| \le -5$ doesn't have a solution.

>> **If the absolute value is greater than or equal to a negative number, there are infinitely many solutions, and the answer is "all real numbers."** An absolute value indicates a positive number or zero, which is always greater than a negative number. For instance, it doesn't matter which number you plug into the equation $|3x + 5| > -2$, you always get a true statement. Therefore, the solution to this statement is "all real numbers."

Following are a couple example questions, as well as a handful of practice problems, to reacquaint you with the process for solving inequalities.

EXAMPLE

Q. Solve for x in $5 - 2x > 4$.

A. $x < \dfrac{1}{2}$

Subtract 5 from each side of $5 - 2x > 4$ to get $-2x > -1$. Next, divide both sides by -2 (don't forget to switch that inequality!). You wind up with $x < \dfrac{1}{2}$.

Q. Solve for x in $|4x + 4| - 3 \geq 9$.

A. $x \geq 2$ or $x \leq -4$

First, you have to isolate the absolute value. To do this, add 3 to both sides: $|4x + 4| \geq 12$. Next, drop the absolute value bars and set up your two inequalities: $4x + 4 \geq 12$ *or* $4x + 4 \leq -12$. Solving each inequality algebraically, you get $x \geq 2$ or $x \leq -4$.

---

1 Solve for x in $|4 - 2x| > 12$.

2 Solve for x in $x^2 - 5x - 20 > 4$.

---

3 Solve for x in $|2x + 16| + 15 > 5$.

4 Solve for x in $x^3 - 5x > 4x^2$.

Expressing Inequality Solutions in Interval Notation

You may or may not have experienced interval notation your earlier math studies. *Interval notation*, although strange sounding, is simply another way of expressing a solution set. Why have another way to write the same thing? Well, this notation is important to know because it's the one most frequently used in pre-calculus and calculus.

TECHNICAL STUFF

The key to writing a solution in interval notation is to locate the beginning and end of a set of solutions. You can do this by starting with inequality notation or by graphing the solution in order to visualize it. After you locate your key points, you need to decide which of the following types of intervals you're dealing with:

>> An **open interval** is always indicated by parentheses in interval notation. For instance, (1, 2) indicates $1 < x < 2$. You show this open interval on a graph by using an open dot at the left endpoint, 1, and another open dot at the right endpoint, 2.

>> A **closed interval** is always indicated by brackets in interval notation. For example, [0, 3] means that $0 \leq x \leq 3$. To graph this closed interval, use a filled dot at the left endpoint, 0, and another filled dot at the right endpoint, 3.

>> A **mixed interval** is indicated by a mix of parentheses and brackets. For instance, both [−4, 3) and (4, 6] are the interval notation for a mixed interval. (Note that $[-4, 3)$ is the interval notation for the solution set $-4 \leq x < 3$, shown in Figure 2-2. Another way to think of this solution set is $x \geq -4$ *and* $x < 3$.) You show a mixed interval on a graph by using either a filled dot at the left endpoint and an open dot at the right endpoint, or an open dot at the left endpoint and a filled dot at the right endpoint.

FIGURE 2-2:
Graph of
$-4 \leq x < 3$.

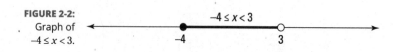

To indicate a solution set that includes non-overlapping sections (also known as *disjoint sets*), you need to state all the intervals of the solution separated by the word *or*. For example, to write the solution set of $x < 2$ or $x \geq 5$ (as shown in Figure 2-3), you need to write both intervals in interval notation: $(-\infty, 2) \cup [5, \infty)$. The symbol between the two sets is the *union* symbol, $\cup$, and it means that the solution can belong in one or the other interval.

FIGURE 2-3:
Graph of $x < 2$
or $x \geq 5$.

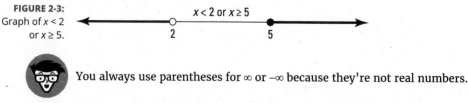

You always use parentheses for ∞ or $-\infty$ because they're not real numbers.

TECHNICAL STUFF

Here are some more examples of dealing with interval notation.

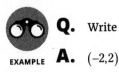

Q. Write the solution for $x^2 < 4$ in interval notation.

EXAMPLE **A.** $(-2, 2)$

The inequality $x^2 < 4$ is true as long as x is smaller than 2 and greater than -2, written $-2 < x < 2$. When writing this using interval notation, you use parentheses.

Q. Graph the interval set $(-2, 3] \cup (5, \infty)$ on a number line.

A. See the graph.

To create your own graph of this interval set, put your key points on a number line and then place the correct open or filled dots on your key points, depending on whether they're closed, open, or mixed intervals. Finally, thicken the interval by shading it.

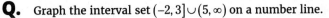

5 Write the solution for $|2x + 16| + 15 > 5$ in interval notation.

6 Write the solution of $x^3 - 5x > 4x^2$ in interval notation.

7 Graph the interval set $(-\infty, -7) \cup [5, 2) \cup (4, \infty)$ on a number line.

8 Graph the solution of $|2x - 1| \le 3$.

Radicals and Exponents — Just Simplify!

REMEMBER

Radicals and exponents (also known as *roots* and *powers*) are fundamental algebra concepts. A *radical* signifies the principal root of a number and is indicated by the radical symbol, $\sqrt{}$. A *square root* of a number is a value that must be multiplied by itself to equal that number. For example, the square root of 9 is 3 because 3 multiplied by itself is 9. However $(-3)(-3)$ is also equal to 9, but when asked to find $\sqrt{9}$, you give the answer $+3$, the positive and principal root. If you're asked to solve $x^2 = 9$, then you include the negative roots. There are other roots; the third root (or *cube root*) of 8 is 2 because 2 multiplied by itself three times is 8: $2 \cdot 2 \cdot 2 = 8$.

TIP

When an exponent is a positive integer, it indicates the number of times a number (the base) is multiplied by itself. For example, 3 to the power of 4 is the same as $3^4 = 3 \cdot 3 \cdot 3 \cdot 3 = 81$.

When an exponent is a rational number, like in the expression $8^{\frac{2}{3}}$, the exponent, $\frac{2}{3}$, isn't equivalent to the number of times 8 is multiplied by itself. Fractional exponents have the root in the denominator and power in the numerator. To evaluate $8^{\frac{2}{3}}$, you can write it as $\left(8^{\frac{1}{3}}\right)^2 = \left(\sqrt[3]{8}\right)^2$ or $\left(8^2\right)^{\frac{1}{3}} = \sqrt[3]{8^2}$.

The denominator of a fractional exponent is the *index* of the radical. The index of the radical is the superscript you see before the radical sign. If the index of a radical is odd, then the value of the radical expression may be negative, but if the index of a radical is even, the result can't be negative.

Roots and exponents are closely related to each other. In fact, they're inverse operations. To solve an equation in which the variable is under a radical, simply raise both sides to the same power. For example, to solve $\sqrt{x} = 4$ you need to square both sides to get $x = 16$. Similarly, you can often solve an equation in which the variable is raised to a power (or has an exponent) by taking the root of both sides. For instance, to solve $x^3 = 27$, you can take the cube root of each side $\sqrt[3]{x^3} = \sqrt[3]{27}$ to get $x = 3$. You can now use these facts to solve equations with radicals and exponents.

TIP

Sometimes it's easier to solve expressions with radicals and exponents by rewriting them as *rational exponents* — exponents written as fractions. To do this, remember that the numerator of the rational exponent (the top number) is the power, and the denominator (the bottom number) is the index of the radical: $x^{\frac{m}{n}} = \sqrt[n]{x^m} = \left(\sqrt[n]{x}\right)^m$.

EXAMPLE

The following example questions show you how to work through equations featuring radicals and exponents before you face the practice problems.

Q. Solve for x in $x^2 - 3x^{3/2} - 4x = 0$.

A. $x = 0, 16$

This problem has all sorts of interesting features. You start by factoring out the x. Doing so leaves you with $x\left(x - 3x^{1/2} - 4\right) = 0$. Now, letting $y = \sqrt{x} = x^{1/2}$, you can recognize that what's in the parentheses is a quadratic with the form: $y^2 - 3y - 4 = 0$. Factor this polynomial: $(y - 4)(y + 1) = 0$. Now you can replace the y terms with the original x and can write $\left(x^{1/2} - 4\right)\left(x^{1/2} + 1\right) = 0$. Next, set each factor of $x\left(x^{1/2} - 4\right)\left(x^{1/2} + 1\right) = 0$ equal to 0 to find your solutions. They should look something like this:

$$x = 0 \qquad \begin{aligned} x^{1/2} - 4 &= 0 \\ x^{1/2} &= 4 \\ \left(x^{1/2}\right)^2 &= 4^2 \\ x &= 16 \end{aligned} \qquad \begin{aligned} x^{1/2} + 1 &= 0 \\ x^{1/2} &= -1 \end{aligned}$$

However, $x^{1/2} + 1 = 0$ has no solution because the principal square root of a number is always non-negative. Hence, $x = 0, 16$. (If you need a refresher of exponential rules, flip to Chapter 5 for a quick review. To remind yourself how to solve quadratics, check out Chapter 4.)

Q. Solve for x in $\sqrt{2x - 1} + 4 = x + 2$.

A. $x = 5$

Start this one by subtracting 4 from both sides of the equation to isolate the radical: $\sqrt{2x - 1} = x - 2$. Next, square each side to get rid of the square root. So $\left(\sqrt{2x - 1}\right)^2 = (x - 2)^2$ becomes $2x - 1 = x^2 - 4x + 4$.

Now you can bring all the terms to the right-hand side, as in $0 = x^2 - 6x + 5$, and then factor to get $0 = (x - 5)(x - 1)$. Setting each factor equal to 0, you get two possible solutions: $x = 5$ or $x = 1$. Plug both solutions back into the original equation to check for extraneous roots (refer to Chapter 1 for more on these). When you plug in your solutions, here's what you find:

$$\sqrt{2(5) - 1} + 4 \overset{?}{=} 5 + 2 \qquad \sqrt{2(1) - 1} + 4 \overset{?}{=} 1 + 2$$
$$\sqrt{10 - 1} + 4 \overset{?}{=} 7 \qquad \sqrt{2 - 1} + 4 \overset{?}{=} 3$$
$$\sqrt{9} \overset{?}{=} 3 \qquad \sqrt{1} \overset{?}{=} -1$$
$$3 = 3 \qquad 1 \neq -1$$

Because these steps are reversible, you see that $x = 1$ is an extraneous root, and $x = 5$ is the solution!

9 Simplify: $27^{4/3}$

 10 Solve for x in $x^{5/3} - 6x = x^{4/3}$.

 11 Solve for x in $\sqrt{x-3} - 5 = 0$.

 12 Solve for x in $x^{8/9} = 16x^{2/9}$.

13 Solve for x in $\sqrt{x-7} - \sqrt{2x-7} = -2$.

14 Solve for x in $x^{2/3} + 7x^{1/3} + 10 = 0$.

Getting Out of a Sticky Situation, or Rationalizing

It may sound irrational, but rationalizing can come in handy sometimes. To simplify fraction involving a radical expression, you often need to rationalize the denominator. In this section, you find a review and practice *rationalizing the denominator*.

First up are *monomials* that need rationalizing. With monomials, you're dealing with an expression and not an equation, so a big hint here is to remember equivalent fractions.

WARNING

Keep in mind that a monomial is an expression, *not* an equation. You can't simply square the term to find a solution because you can't counterbalance that action. Instead, you need to multiply the numerator and denominator by the same term (which is the same as multiplying by 1). For example, if you need to rationalize the expression $\frac{3}{\sqrt{2}}$, you can multiply the expression by $\frac{\sqrt{2}}{\sqrt{2}}$, which equals 1. You then get $\frac{3}{\sqrt{2}} \cdot \frac{\sqrt{2}}{\sqrt{2}} = \frac{3\sqrt{2}}{2}$.

The same idea works for other radicals with other indexes, but what you multiply by has to create a perfect power under the radical in the denominator. For example, if you need to rationalize the expression $\frac{2}{\sqrt[3]{5}}$, you need to multiply the numerator and denominator by $\sqrt[3]{5}$ to the second power, or by $\sqrt[3]{5^2}$, because you want to find the cube root of a third power. After multiplying, you get $\frac{2}{\sqrt[3]{5}} \cdot \frac{\sqrt[3]{5^2}}{\sqrt[3]{5^2}} = \frac{2\sqrt[3]{25}}{\sqrt[3]{125}} = \frac{2\sqrt[3]{25}}{5}$.

To rationalize expressions with binomials in the denominator, you must multiply both the numerator and denominator by the conjugate of that denominator. A *conjugate* is a fancy name for the binomial that gives you the difference of two squares when multiplied by the first binomial. It's found by changing the sign of the second term of the binomial. For example, the conjugate of $x + y$ is $x - y$ because when you multiply the two conjugates $(x + y)(x - y)$, you get $x^2 - y^2$, or the difference of two squares. So to rationalize a denominator with a binomial, multiply the numerator and denominator by the conjugate and then simplify. For example, to simplify $\frac{3}{2 - \sqrt{3}}$, multiply the numerator and denominator by $2 + \sqrt{3}$. The steps look like this:

$$\frac{3}{2 - \sqrt{3}} \cdot \frac{2 + \sqrt{3}}{2 + \sqrt{3}} = \frac{3(2 + \sqrt{3})}{(2 - \sqrt{3})(2 + \sqrt{3})} = \frac{3(2 + \sqrt{3})}{4 - 3} = \frac{3(2 + \sqrt{3})}{1} = 6 + 3\sqrt{3}$$

Q. Simplify: $\dfrac{12}{\sqrt[4]{9}}$

A. $4\sqrt[4]{9}$

First things first: Rewrite the denominator to get $\dfrac{12}{\sqrt[4]{3^2}}$. Then multiply the numerator and denominator by $\sqrt[4]{3^2}$ to have a fourth power under the radical. $\dfrac{12}{\sqrt[4]{3^2}} \cdot \dfrac{\sqrt[4]{3^2}}{\sqrt[4]{3^2}} = \dfrac{12\sqrt[4]{3^2}}{\sqrt[4]{3^4}}$

$= \dfrac{\cancel{12}\,\sqrt[4]{9}}{\cancel{3}} = 4\sqrt[4]{9}$

Q. Simplify: $\dfrac{2+\sqrt{5}}{3-\sqrt{5}}$

A. $\dfrac{11+5\sqrt{5}}{4}$

Multiply the numerator and denominator by the conjugate of the denominator:

$\dfrac{2+\sqrt{5}}{3-\sqrt{5}} \cdot \dfrac{3+\sqrt{5}}{3+\sqrt{5}}$

Multiply out the expression by using FOIL:

$$\dfrac{2+\sqrt{5}}{3-\sqrt{5}} \cdot \dfrac{3+\sqrt{5}}{3+\sqrt{5}} = \dfrac{\left(2+\sqrt{5}\right)\left(3+\sqrt{5}\right)}{\left(3-\sqrt{5}\right)\left(3+\sqrt{5}\right)} = \dfrac{6+2\sqrt{5}+3\sqrt{5}+5}{9-5} = \dfrac{11+5\sqrt{5}}{4}$$

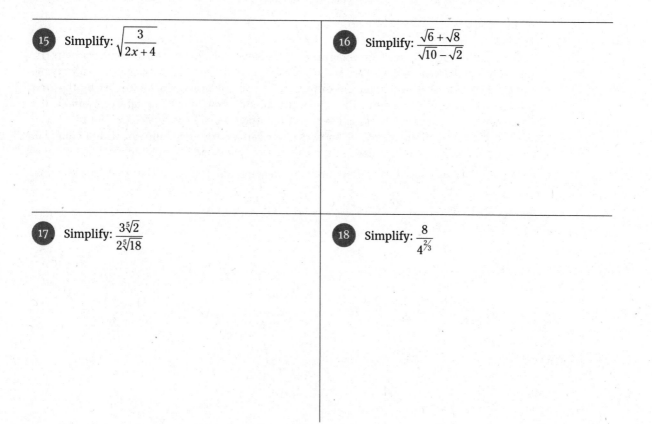

15 Simplify: $\sqrt{\dfrac{3}{2x+4}}$

16 Simplify: $\dfrac{\sqrt{6}+\sqrt{8}}{\sqrt{10}-\sqrt{2}}$

17 Simplify: $\dfrac{3\sqrt[5]{2}}{2\sqrt[5]{18}}$

18 Simplify: $\dfrac{8}{4^{\frac{2}{3}}}$

Answers to Problems on Real Numbers

This section contains the answers for the practice problems presented in this chapter. You'll also find explanations for the answers.

(1) Solve for x in $|4-2x|>12$. The answer is $x>8$ or $x<-4$.

Start by dropping the absolute value bars and setting up your two inequalities: $4-2x>12$ or $4-2x<-12$. Solve algebraically (be careful when you divide by that negative, though): $-2x>8$ becomes $x<-4$ and $-2x<-16$ becomes $x>8$.

(2) Solve for x in $x^2-5x-20>4$. The answer is $x>8$ or $x<-3$.

First, you need to recognize that you're dealing with a quadratic and recall that to solve a quadratic, you need to isolate its terms on one side of the inequality (for a refresher on quadratics, see Chapter 4). Subtract 4 from both sides of the equation: $x^2-5x-24>0$. Next, factor your quadratic: $(x-8)(x+3)>0$. Setting each factor to 0 gives you your key points: 8 and –3. If you put these on a number line, you can see that you have three possible solutions: less than –3, between –3 and 8, and greater than 8. All you have to do is plug in numbers in each interval to see whether you get a positive or negative number. Because you're looking for a solution that's greater than 0, you need a positive result when you multiply your factors, $(x-8)$ and $(x+3)$, together. In other words, you want both of your factors to be positive or both of them to be negative. Looking at the number line, you see that your solutions are $x>8$ or $x<-3$.

(3) Solve for x in $|2x+16|+15>5$. The answer is "all real numbers."

Subtracting 15 from both sides in order to isolate the absolute value gives you $|2x+16|>-10$. Because you know that absolute values are positive or zero and therefore greater than any negative, you can conclude that the solution is "all real numbers."

(4) Solve for x in $x^3-5x>4x^2$. The answer is $-1<x<0$ or $x>5$.

To solve this problem, gather all your variables to one side of the equation by subtracting $4x^2$ from both sides: $x^3-4x^2-5x>0$. Next, factor out x from each term: $x(x^2-4x-5)>0$. Then factor the quadratic: $x(x-5)(x+1)>0$. Set your factors equal to 0 so you can find your key points. When you have them, put these points on a number line and plug in test numbers from each possible section to determine whether the factor would be positive or negative.

$x(x-5)(x+1)$	$x(x-5)(x+1)$	$x(x-5)(x+1)$	$x(x-5)(x+1)$
$-(-)(-)$	$-(-)(+)$	$+(-)(+)$	$+(+)(+)$

$$\longleftarrow \!\!\!\!\!\!\underset{-1}{\circ} \!\!\!\!\!\!\underset{0}{\circ} \!\!\!\!\!\!\underset{5}{\circ}\!\!\!\!\!\!\longrightarrow$$

Doing the multiplication, and looking for an even number of negative signs, you see that your solution is $-1<x<0$ or $x>5$.

(5) Write the solution for $|2x+16|+15>5$ in interval notation. The answer is $(-\infty,\infty)$

In problem 3, you determine that the solution is all real numbers, and you write that in interval notation by writing it as infinity to negative infinity.

(6) Write the solution of $x^3 - 5x > 4x^2$ in interval notation. The answer is $(-1, 0), (5, \infty)$.

In problem 4, you determine that the solution is $-1 < x < 0$ or $x > 5$.

(7) Graph the interval set $(-\infty, -7) \cup [5, 2) \cup (4, \infty)$ on a number line. See the graph.

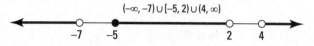

$(-\infty, -7) \cup [-5, 2) \cup (4, \infty)$

$-7 \quad -5 \qquad\qquad 2 \quad 4$

(8) Graph the solution of $|2x - 1| \le 3$. See the graph.

Start by dropping the absolute value sign and setting up your two inequalities: $2x - 1 \le 3$ and $2x - 1 \ge -3$. Then solve both inequalities to find your solution: $x \le 2$ and $x \ge -1$. You can rewrite these solutions as $-1 \le x \le 2$, which you can then graph.

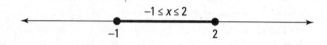

$-1 \le x \le 2$

$-1 \qquad\qquad 2$

(9) Simplify $27^{4/3}$. The answer is 81.

You can think of this problem in two ways: $\sqrt[3]{27^4}$ and $\left(\sqrt[3]{27}\right)^4$. Either way gives you the correct answer, but one is easier to deal with than the other.

Starting with $\sqrt[3]{27^4}$, the order of operations tells you to raise 27 to the power of 4 first. Doing so gives you $\sqrt[3]{531,441}$ — an awfully big number. Now you have to find the cube root.

Solving this problem is much easier when you choose to deal with it written like this: $\left(\sqrt[3]{27}\right)^4$. In this case, the order of operations tells you to take the cube root of 27, which is 3, and then raise 3 to the fourth power, which is 81. Ah . . . much better.

(10) Solve for x in $x^{5/3} - 6x = x^{4/3}$. The answer is $x = 0, -8, 27$.

Your first task is to bring all the terms to one side in descending order: $x^{5/3} - x^{4/3} - 6x = 0$. Next, factor out an x from each term: $x\left(x^{2/3} - x^{1/3} - 6\right) = 0$. The resulting expression is similar to $y\left(y^2 - y - 6\right) = 0$, which factors into $y(y + 2)(y - 3) = 0$. Similarly, you can factor $x\left(x^{2/3} - x^{1/3} - 6\right) = 0$ into $x\left(x^{1/3} + 2\right)\left(x^{1/3} - 3\right) = 0$, and then set each factor equal to 0 and solve.

Here's the solution:

$$x = 0$$

$$x^{1/3} + 2 = 0 \qquad\qquad x^{1/3} - 3 = 0$$
$$x^{1/3} = -2 \qquad\qquad x^{1/3} = 3$$
$$\left(x^{1/3}\right)^3 = (-2)^3 \qquad\qquad \left(x^{1/3}\right)^3 = (3)^3$$
$$x = -8 \qquad\qquad x = 27$$

(11) Solve for x in $\sqrt{x - 3} - 5 = 0$. The answer is $x = 28$.

Isolate the radical by adding 5 to each side of the equation to get $\sqrt{x - 3} = 5$. Then square both sides to get rid of the square root, like so: $\left(\sqrt{x - 3}\right)^2 = 5^2$. This equation simplifies to $x - 3 = 25$, which is the same as $x = 28$. Checking to be sure it isn't extraneous shows that it is the solution.

(12) Solve for x in $x^{8/9} = 16x^{2/9}$. The answer is $x = 0, -64, 64$.

Bring both terms to the same side: $x^{8/9} - 16x^{2/9} = 0$.

Next, factor out $x^{2/9}$ from each term to get $x^{2/9}\left(x^{6/9} - 16\right) = 0$ and simplify the exponent in the parentheses to get $x^{2/9}\left(x^{2/3} - 16\right) = 0$. You can see that the expression in parentheses has the same form as $y^2 - 16$, which factors into $(y+4)(y-4) = 0$. Similarly, you can factor $x^{2/9}\left(x^{2/3} - 16\right) = 0$ into $x^{2/9}\left(x^{1/3} + 4\right)\left(x^{1/3} - 4\right) = 0$.

Last but not least, set each factor equal to 0 so you can find your solutions:

$$x^{2/9} = 0 \qquad\qquad x^{1/3} + 4 = 0 \qquad\qquad x^{1/3} - 4 = 0$$
$$\left(x^{2/9}\right)^{9/2} = 0^{9/2} \qquad x^{1/3} = -4 \qquad\qquad x^{1/3} = 4$$
$$x = 0 \qquad\qquad \left(x^{1/3}\right)^3 = (-4)^3 \qquad \left(x^{1/3}\right)^3 = (4)^3$$
$$x = -64 \qquad\qquad x = 64$$

(13) Solve for x in $\sqrt{x-7} - \sqrt{2x-7} = -2$. The answer is $x = 8, 16$.

Begin by isolating one of the radicals: $\sqrt{x-7} = \sqrt{2x-7} - 2$. Then, square both sides to get rid of that radical, $\left(\sqrt{x-7}\right)^2 = \left(\sqrt{2x-7} - 2\right)^2$, and make sure you multiply your binomials correctly: $\left(\sqrt{x-7}\right)^2 = \left(\sqrt{2x-7} - 2\right)\left(\sqrt{2x-7} - 2\right)$. Squaring on the left and multiplying the terms on the right side of the equation gives you $x - 7 = 2x - 7 - 4\sqrt{2x-7} + 4$. Next, isolate the remaining radical: $4\sqrt{2x-7} = x + 4$.

Then, you can square both sides again to remove the remaining radical: $\left(4\sqrt{2x-7}\right)^2 = (x+4)^2$.

Again, square on the left and multiply the two binomials on the right and combine like terms: $16(2x-7) = x^2 + 8x + 16$ becomes $32x - 112 = x^2 + 8x + 16$. Moving all the terms to the right, you get $0 = x^2 - 24x - 128$. This quadratic factors into $0 = (x-8)(x-16)$. When you set both factors equal to 0, you get two possible solutions: $x = 8$ and $x = 16$. Plug both back into the original equation to find that both solutions work.

(14) Solve for x in $x^{2/3} + 7x^{1/3} + 10 = 0$. The answer is $x = -8, -125$.

Start by recognizing that this trinomial is similar to $y^2 + 7y + 10 = 0$, which factors to $(y+5)(y+2) = 0$.

Similarly, $x^{2/3} + 7x^{1/3} + 10 = 0$ factors into $\left(x^{1/3} + 5\right)\left(x^{1/3} + 2\right) = 0$. By setting each factor equal to 0, you can easily solve for the solutions. Just raise each side to the power of 3. In other words, $x^{1/3} + 5 = 0$ becomes $\left(x^{1/3}\right)^3 = (-5)^3$, so $x = -125$. Similarly, $x^{1/3} + 2 = 0$ becomes $\left(x^{1/3}\right)^3 = (-2)^3$, so $x = -8$.

(15) Simplify $\sqrt{\dfrac{3}{2x+4}}$. The answer is $\dfrac{\sqrt{6x+12}}{2x+4}$.

First, you need to separate the fraction into two radicals: one in the numerator and one in the denominator: $\dfrac{\sqrt{3}}{\sqrt{2x+4}}$. Next, multiply the numerator and denominator by the square root in the denominator: $\dfrac{\sqrt{3}}{\sqrt{2x+4}} \cdot \dfrac{\sqrt{2x+4}}{\sqrt{2x+4}}$. This one doesn't require the use of a conjugate because there isn't another term added to the radical. Just simplify the numerator by multiplying the radicals: $\dfrac{\sqrt{3}}{\sqrt{2x+4}} \cdot \dfrac{\sqrt{2x+4}}{\sqrt{2x+4}} = \dfrac{\sqrt{3}\sqrt{2x+4}}{2x+4} = \dfrac{\sqrt{6x+12}}{2x+4}$.

(16) Simplify $\dfrac{\sqrt{6}+\sqrt{8}}{\sqrt{10}-\sqrt{2}}$. The answer is $\dfrac{\sqrt{15}+\sqrt{3}+2\sqrt{5}+2}{4}$.

Multiply the numerator and denominator by the conjugate of the denominator to get $\dfrac{\sqrt{6}+\sqrt{8}}{\sqrt{10}-\sqrt{2}} \cdot \dfrac{\sqrt{10}+\sqrt{2}}{\sqrt{10}+\sqrt{2}} = \dfrac{\left(\sqrt{6}+\sqrt{8}\right)\left(\sqrt{10}+\sqrt{2}\right)}{10-2}$ and then use FOIL to multiply the binomials in the numerator and denominator; the result is $\dfrac{\sqrt{60}+\sqrt{12}+\sqrt{80}+\sqrt{16}}{8}$. Simplify each radical: $\dfrac{\sqrt{60}+\sqrt{12}+\sqrt{80}+\sqrt{16}}{8} = \dfrac{2\sqrt{15}+2\sqrt{3}+4\sqrt{5}+4}{8}$. Finally, because each term in the numerator and denominator is divisible by 2, divide all terms by 2: $= \dfrac{\sqrt{15}+\sqrt{3}+2\sqrt{5}+2}{4}$.

(17) Simplify $\dfrac{3\sqrt[5]{2}}{2\sqrt[5]{18}}$. The answer is $\dfrac{\sqrt[5]{3^3}}{2}$.

Begin by factoring the denominator: $\dfrac{3\sqrt[5]{2}}{2\sqrt[5]{18}} = \dfrac{3\sqrt[5]{2}}{2\sqrt[5]{2\cdot 9}} = \dfrac{3\sqrt[5]{2}}{2\sqrt[5]{2}\sqrt[5]{9}}$. Notice the $\sqrt[5]{2}$ in both the numerator and denominator? Reduce the fraction: $\dfrac{3\cancel{\sqrt[5]{2}}}{2\cancel{\sqrt[5]{2}}\sqrt[5]{9}} = \dfrac{3}{2\sqrt[5]{9}} = \dfrac{3}{2\sqrt[5]{3^2}}$. Next, multiply the numerator and denominator by $\sqrt[5]{3^3}$ to eliminate the radical in the denominator: $\dfrac{3}{2\sqrt[5]{3^2}} \cdot \dfrac{\sqrt[5]{3^3}}{\sqrt[5]{3^3}} = \dfrac{3\sqrt[5]{3^3}}{2\sqrt[5]{3^5}} = \dfrac{3\sqrt[5]{3^3}}{2\cdot 3}$. Now reduce the fraction to arrive at your final answer, which is $\dfrac{\cancel{3}\sqrt[5]{3^3}}{2\cdot\cancel{3}} = \dfrac{\sqrt[5]{3^3}}{2}$.

(18) Simplify $\dfrac{8}{4^{2/3}}$. The answer is $2\sqrt[3]{4}$.

Change the fractional exponent into a radical to get $\sqrt[3]{4^2}$ and then multiply the numerator and denominator by one more cube root of 4: $\dfrac{8}{\sqrt[3]{4^2}} \cdot \dfrac{\sqrt[3]{4}}{\sqrt[3]{4}} = \dfrac{8\sqrt[3]{4}}{\sqrt[3]{4^3}}$. Simplify to get $\dfrac{8\sqrt[3]{4}}{\sqrt[3]{4^3}} = \dfrac{\overset{2}{\cancel{8}}\sqrt[3]{4}}{\cancel{4}} = 2\sqrt[3]{4}$.

Chapter **3**

Controlling Functions by Knowing Their Function

By this point in your math studies, you're familiar with the coordinate plane (it's what you get when two number lines meet at a 90-degree angle). You know that the horizontal axis is called the x-axis and that the vertical one is called the y-axis. You also know that each point, or ordered pair, on the plane is named (x, y). But did you know that a set of ordered pairs is called a relation? The domain of the relation is the set of all the x values, and the range is the set of all the y values. For convenience, the domain variable always precedes the range variable in order and alphabetically in this book.

A *function* is a relation where every x in the domain pairs with one (and only one) y in the range. Think of a function as a computer. Domain is input, and range is output. A common symbol denoting a function is $f(x)$, and it reads "f of x." You find the concept of functions, as well as some of their properties, in this chapter.

Using Both Faces of the Coin: Even and Odd

If you've ever taken an art class, you've probably heard the term *symmetry*. It means that what you're seeing is balanced, with equal or similar parts on both sides of the object, be it a painting, sculpture, or photograph. A graph can be symmetrical as well. The graph of a relation can have any of the following three basic types of symmetry:

» **Y-axis symmetry:** Each point on the left side of the *y*-axis is mirrored by a point on the right side (and vice versa).

» **X-axis symmetry:** Each point above the *x*-axis is mirrored by a point below it (and vice versa).

» **Origin symmetry:** A graph has this type of symmetry if it's unchanged when reflected across both the *x*-axis and *y*-axis. In other words, if you turn a graph with origin symmetry upside down and then flip it left to right, it looks exactly the same as it did before the flips.

In pre-calculus, functions use the idea of symmetry to describe special properties.

TECHNICAL STUFF

» **Even function:** This is a function whose graph is symmetrical with respect to the *y*-axis. Basically, each input *x* and the opposite input $-x$ give the same *y* value, which means $f(x) = f(-x)$.

» **Odd function:** This is any function whose graph is symmetrical with respect to the origin. This means that each *x* value gives a *y* value, and its opposite $-x$ gives the opposite $-y$, which means that $f(-x) = -f(x)$.

Q. Determine whether $f(x) = x^4 - x^2$ is even, odd, or neither.

EXAMPLE **A.** This function is even.

Replace *x* with $-x$ in the function rule and see what happens: $f(-x) = (-x)^4 - (-x)^2$. A negative number raised to an even power is a positive number, so $f(-x) = x^4 - x^2$. Because you get the same exact function as the original one, this function is even.

 Is $f(x) = x^3 - 1$ even, odd, or neither?

2 Does the given graph appear to be even, odd, or neither?

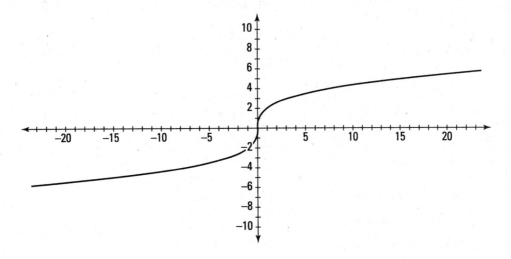

---

3 Sketch half the graph (the half where x is positive) $f(x) = \sqrt{x^2 - 4}$ and use symmetry to complete the graph.

4 Sketch half the graph of $f(x) = 4x^3$ and use symmetry to complete the graph.

Leaving the Nest: Transforming Parent Graphs

You see certain functions over and over again in pre-calculus, and sometimes you're even asked to graph them. To save you time and give you more confidence in the graphs you're producing, it's recommended that you memorize a few basic graphs of the more commonly used functions, called *parent graphs.*

TECHNICAL STUFF

Common parent graphs include quadratic functions, square roots, absolute values, cubics, and cube roots. Moving these basic graphs around the coordinate plane is known as *transforming the function* and is easier than using the plug-and-chug method. Several types of function transformations exist:

>> Horizontal stretches

>> Vertical flattenings

>> Reflections

>> Horizontal translations

>> Vertical translations

In the sections that follow, you'll find several parent functions and then how to transform them. *Note:* Even though in most sections you'll see only one function when discussing the transformations, the rules presented apply to *all* parent functions in the same way. So if the discussion is about a quadratic function in the section on vertical translations, that's not the only function that has vertical translations — they all do.

Quadratic functions

Quadratic functions are defined by second-degree *polynomials* (see Chapter 4) where the highest exponent on any one input variable is 2. The parent quadratic function is $f(x) = x^2$, and its graph is known as a *parabola* (a type of conic section; head to Chapter 12 for the scoop on these). Furthermore, $f(x) = x^2$ is an even function, so the graph of the parent quadratic function is symmetric with respect to the y-axis.

Begin the graph at the *vertex*, which in the case of this function is the *origin*; here the vertex is the point (0, 0). Use the function rule to get a few points on the graph. Sticking to positive input values, you may get (1, 1), (2, 4), (3, 9), and so on. Then you can use symmetry to complete the graph. Check out Figure 3-1 for the graph of the parent quadratic function.

Square root functions

Not surprisingly, *square root functions* feature square roots. The parent square root function is $f(x) = \sqrt{x}$. As you can see in Figure 3-2, this graph looks like half a parabola that's turned on its side. Because the square root of a negative number isn't a real number, you can't have x values that are negative, which is why the parent graph of the square root function doesn't cross into the left side of the coordinate plane. (For more on the square roots of negative numbers, flip to Chapter 11.) This graph begins at the origin (0, 0) and then moves up through the points (1, 1), (4, 2), (9, 3), and so on.

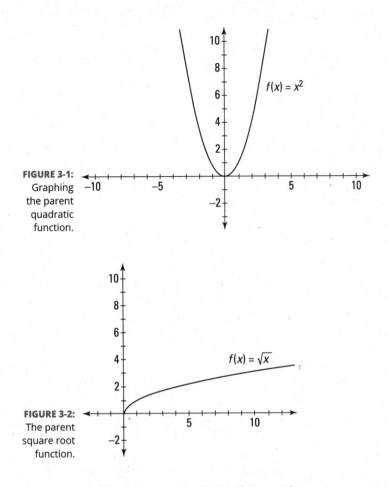

FIGURE 3-1:
Graphing
the parent
quadratic
function.

$f(x) = x^2$

FIGURE 3-2:
The parent
square root
function.

$f(x) = \sqrt{x}$

Absolute value functions

The parent *absolute value function* is $f(x) = |x|$. You should recognize the absolute value bars and know that this function always gives the distance from the origin, so it always gives a non-negative output. Furthermore, it is an even function, so the graph of the parent absolute value function is symmetric with respect to the y-axis.

To graph this function, begin at the vertex (0, 0) and move right to (1, 1). From there, still working on the right-hand side of the coordinate plane, move through the points (2, 2), (3, 3), and so on. Continue this pattern to complete the right side of the graph; use symmetry to complete the left half of the graph (see Figure 3-3).

Cubic functions

Cubic functions are defined by third-degree polynomials, so the highest exponent on any one variable is 3. The parent cubic function is $f(x) = x^3$, which is an odd function. If you turn its graph upside down and flip from left to right (see Figure 3-4), it looks exactly the same. If you start the graph at the origin (0, 0), a point on the graph to the left of (0, 0) is (−1, 0), and a point on the graph to the right of (0, 0) is (1, 1).

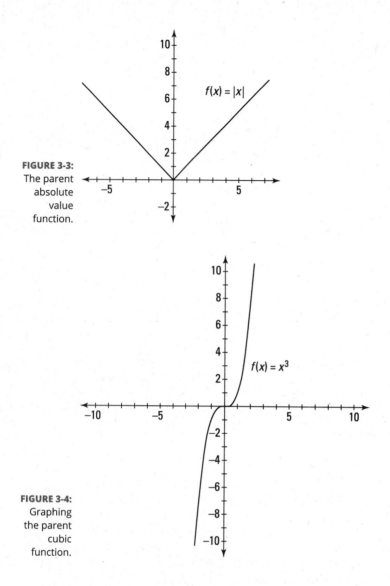

FIGURE 3-3:
The parent
absolute
value
function.

$f(x) = |x|$

$f(x) = x^3$

FIGURE 3-4:
Graphing
the parent
cubic
function.

Cube root functions

Cube root functions are related to cubic functions similar to the way that quadratic and square root functions are related. The parent cube root function is $f(x) = \sqrt[3]{x}$. Like the cubic function (see the preceding section), from the origin, a point to the left on the graph is at $(-1, 1)$, and a point to the right on the graph is $(1, 1)$. However, the graph of the parent cube root function is longer than it is tall. Take a look at Figure 3-5 to see for yourself.

Steeper or flatter

Function graphs can be altered in many ways. One general category involves stretching or steepening versus flattening out the graph.

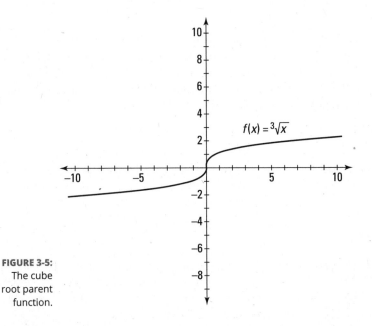

FIGURE 3-5:
The cube root parent function.

REMEMBER

You create a *vertical stretch or flattening* when you multiply a function by a constant, because doing so changes the graph of the function in the vertical direction. This process is written as $a \cdot f(x)$. Think of the result as a vertical stretch or steepening or a vertical shrinking or flattening. A coefficient of a greater than 1 yields a stretch, and a coefficient of a between 0 and 1 yields a shrink. For example, $f(x) = 2x^2$ multiplies each y-value by 2. So from the vertex, instead of moving over 1 and up 1, you now move over 1 and up 2. Every output value from the parent function is multiplied by 2, so it's twice as steep as the original function. You get the idea. For another example, if $g(x) = \dfrac{1}{4}x^2$, then every output value from the parent function is multiplied by $\dfrac{1}{4}$. The graphs of $f(x)$ and $g(x)$ are shown in Figures 3-6a and 3-6b, respectively.

Note: A negative coefficient is actually a reflection; that type of transformation is covered in the later "Reflections" section.

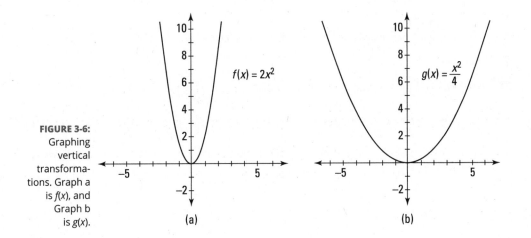

FIGURE 3-6:
Graphing vertical transformations. Graph a is $f(x)$, and Graph b is $g(x)$.

Translations

An action that moves a graph horizontally or vertically (without changing its shape) on the coordinate plane is called a *translation*. The result of a translation is that every point on the parent graph is moved right, left, up, or down by the same amount. In the next sections, you get a closer look at the two types of translations, or *shifts*.

Vertical shifts

Adding a number to or subtracting a number from a given function is a *vertical shift* and is written as $f(x)+c$, where c is the vertical shift. For example, $p(x)=x^3-1$ moves the parent cubic function down by 1, and $p(x)=x^3+4$ moves the parent cubic function up by 4.

Horizontal shifts

Adding a number to or subtracting a number from the function's independent variable (input) is a *horizontal shift*. This type of shift is always written in the form $f(x-c)$. For example, $h(x)=\sqrt{x-2}$ moves the parent square root function to the right by 2, and $h(x)=\sqrt{x+3}$ moves the parent square root function to the left by 3.

Reflections

Reflections take the parent function and reflect it over a horizontal or vertical line. A reflection of a parent function the x-axis takes the form $-f(x)$. In this case, the y-values are all negated. A reflection over the y-axis has the form $f(-x)$ where each x value is switched.

Combinations of transformations

Putting some or all of the various transformations into one function is itself a transformation. Here's what a transformation involving some of the various possibilities looks like: $a \cdot f\left[c(x-h)\right]+k$ where

>> a is the stretch or shrinkage; when a is negative you have a reflection

>> c affects stretch or shrinkage; when c is negative you have a reflection

>> h is the horizontal translation

>> k is the vertical translation

TIP

When graphing, it's best to do the translations first and then doing the other transformations.

Following are two examples of how to transform the parent graph of a function, followed by six practice questions.

Q. Graph the function $f(x) = (x-3)^2$ by transforming the parent graph.

EXAMPLE **A.** See the graph. This transformation is done in one step. Because the constant is subtracted inside the quadratic function, you recognize it as a horizontal shift to the right by 3. Take the parent quadratic function and move each point to the right by 3, as shown in the graph.

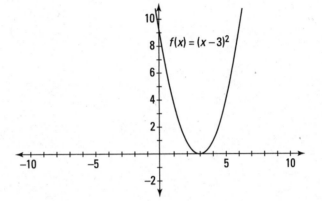

Q. Sketch the graph of $g(x) = \sqrt{3-x} + 1$ by transforming the parent function.

A. See the graph. This one takes some work before you can begin graphing it. You first have to rewrite it in the proper form before you can recognize the various transformations to the parent square root function. Rewrite the terms inside the square root first: $g(x) = \sqrt{-(x-3)} + 1$. There are three transformations being performed on the parent function square root: translation to the right by 3, translation down by 1, and a vertical reflection. Do the translations first, and then reflect over the vertical line $x = 3$.

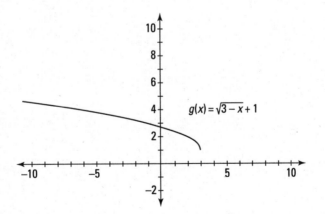

Graph the function $a(x) = -2(x-1)^2$.

 Graph the function $b(x) = |x+4| - 1$.

 Graph the function $c(x) = \sqrt{x+3}$.

 Graph the function $f(x) = -x^2 - 6x$.

Given the graph of the function g(x) in the following figure, sketch the graph of the functions in Questions 9 and 10.

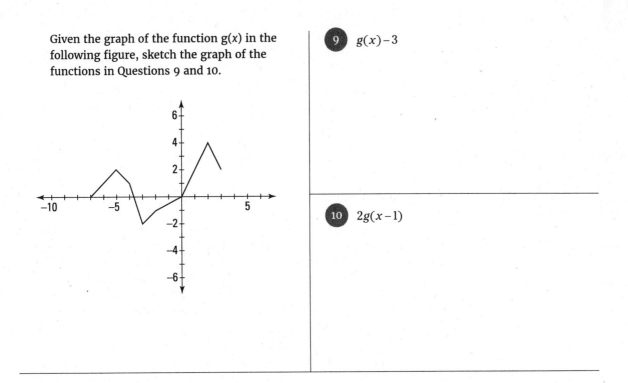

9 $g(x)-3$

10 $2g(x-1)$

Graphing Rational Functions

A *rational function*, such as $f(x)=\dfrac{x+1}{(x-3)^2}$, is the quotient of two polynomials, excluding division by zero (polynomials are covered in Chapter 4). By this point in your math studies, you know that when the denominator of a fraction is 0, the result is undefined. The same is true for rational functions. Because the denominator of a rational function has a variable, it's possible that certain values of the variable will make the denominator 0. Such is the case of $f(x)$, when $x=3$. When a rational function has values that make it undefined, the graph of that function may feature a *vertical asymptote,* which is denoted by a vertical line that the graph never crosses.

To find the vertical asymptote, if there is one, first simplify the rational function to ensure that numerator and denominator don't have any common factors other than constants (no value will simultaneously make the numerator and denominator polynomials zero); then set the denominator equal to 0 and solve it. These solutions (roots) identify locations for vertical asymptotes.

Some rational functions also have a *horizontal asymptote,* which is a horizontal line that a graph approaches as the function input values go toward positive or negative infinity. To find the horizontal asymptote, take a look at the degree of the numerator and the degree of the denominator. (If you don't recall how to find the degree of a polynomial, head to Chapter 4.) Here are the three possibilities for horizontal asymptotes:

TECHNICAL STUFF

>> **The degree of the denominator is greater:** The bottom of the fraction is getting bigger faster, and the fraction goes to 0 as $|x|$ gets larger. Your horizontal asymptote is the x-axis, or $y=0$.

» **The degree of both is the same:** The top and bottom of the fraction are increasing or decreasing at the same rate. The quotient of the lead coefficients gives you the equation of the horizontal asymptote, $y = h$.

» **The degree of the numerator exceeds the degree of the denominator:** The top of the fraction is getting bigger or smaller faster. In short, as x gets larger, so does y; therefore, there's no horizontal asymptote.

Work through these example questions before trying your hand at graphing rational functions.

Q. Graph the function $f(x) = \dfrac{3x-1}{4-x}$.

A. See the graph. First, find the vertical asymptote (if there is one) by setting the denominator equal to 0 and solving. If $4 - x = 0$, then $x = 4$ and the numerator is not zero when $x = 4$. Draw a coordinate plane and add in a dotted vertical line at $x = 4$ to indicate your vertical asymptote. Now, look at the numerator and the denominator; the degree on each is one. Divide the leading coefficients to find the horizontal asymptote. In this case, the numerator's leading coefficient is 3, and the denominator's leading coefficient is –1. This means your horizontal asymptote is $y = \dfrac{3}{-1} = -3$.

Now that you have both asymptotes, use them to help you draw the graph. The vertical asymptote divides your domain into two intervals: $(-\infty, 4)$ and $(4, \infty)$. Pick a couple of x-values on each interval and plug them into the function to determine whether the graph lives above or below the horizontal asymptote. For example, if $x = -5$, then $y = -1.77$. And if $x = 0$, then $y = -0.25$. If you graph those two points, you see that they're both above the horizontal asymptote. Keep checking points such as x and y intercepts and looking at x values close to the vertical asymptote until you have a good idea of what the graph looks like.

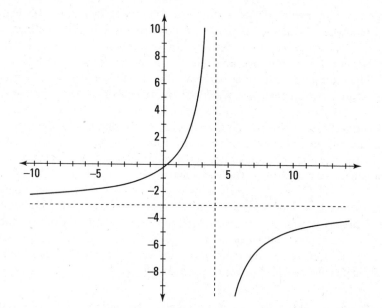

Q. Graph the function $g(x) = \dfrac{2x-6}{x^2+4}$.

A. See the graph. When you try to find a vertical asymptote, you notice that $x^2 + 4 = 0$ doesn't have a solution because $x^2 = -4$ has no solution (in real numbers anyway). Also, because the denominator has a larger degree than the numerator, the horizontal asymptote is the x-axis, or $y = 0$. However, by setting the numerator equal to 0, you do get a solution: $2x - 6 = 0$; $2x = 6$; $x = 3$. This means the graph crosses the x-axis at $x = 3$. Because there's no vertical asymptote, use $x = 3$ to give you the intervals to look at to get the graph.

On the first interval $(-\infty, 3)$, y is negative, and the whole graph is below the horizontal asymptote. On the next interval $(3, \infty)$, y happens to be positive, and the function is above the horizontal asymptote. If you pick x values bigger than 3 that keep getting bigger, you see y increase slowly and then decrease again and get closer and closer to 0. This gives you the graph of this function.

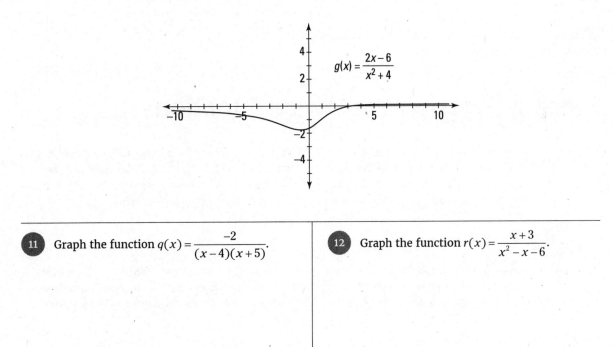

11 Graph the function $q(x) = \dfrac{-2}{(x-4)(x+5)}$.

12 Graph the function $r(x) = \dfrac{x+3}{x^2-x-6}$.

13 Graph the function $t(x) = \dfrac{x^2 - 5x}{x^2 - 4x - 21}$.

14 Graph the function $u(x) = \dfrac{x^2 - 10x - 24}{x + 1}$.

Piecing Together Piecewise Functions

A *piecewise function* is broken into two or more parts. In other words, it actually contains several functions, each of which is defined on a restricted interval. The output depends on what the input is. The graphs of these functions may look like they've been broken into pieces. Because of this broken quality, a piecewise function that jumps is called *discontinuous*.

 Q. Graph $f(x) = \begin{cases} x^2 + 2 & \text{if } x \le 1 \\ 3x - 1 & \text{if } x > -1 \end{cases}$

EXAMPLE **A.** See the graph. This function has been broken into two pieces: When $x \le -1$, the function follows the graph of the quadratic function, and when $x > 1$, the function follows the graph of the linear function. (Notice the hole in this second piece of the graph to indicate that the point isn't actually there.)

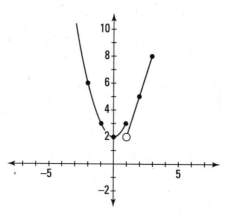

15 Graph $g(x) = \begin{cases} \sqrt{x+3} & \text{if } x \le -1 \\ (x+3)^2 & \text{if } x > -1 \end{cases}$

16 Graph $h(x) = \begin{cases} \dfrac{1}{2}x - 4 & \text{if } x \le -2 \\ 3x + 3 & \text{if } -2 < x < 2 \\ 4 - x & \text{if } x \ge 2 \end{cases}$

17 Graph $m(x) = \begin{cases} x^3 + 2 & \text{if } x < 0 \\ x^2 + 2 & \text{if } 0 \le x < 2 \\ x + 2 & \text{if } x \ge 2 \end{cases}$

18 Graph $n(x) = \begin{cases} |x - 1| & \text{if } x < -3 \\ -3 & \text{if } x = -3 \\ |x| - 1 & \text{if } x > -3 \end{cases}$

Combining Functions

You've come to know, through years and years of practice, the four basic operations in math: addition, subtraction, multiplication, and division. Well, in pre-calculus, you can add, subtract, multiply, and divide them in order to create brand-new functions — a process that's sometimes called *combining functions*.

If you're asked to graph a combined function without using a graphing calculator, you have to plug and chug your way through it by picking plenty of x-values to make sure you get an accurate representation of the graph. You may also be asked to find one specific value for a combined function — you get an x value and you just plug it in and see what happens.

For all the example and practice questions that follow in this section, you use three functions:

> $f(x) = x^2 - 6x + 2$

> $g(x) = 2x^2 - 5x$

> $h(x) = \sqrt{3x + 2}$

Q. Find $(f - g)x$.

A. $(f - g)(x) = -x^2 - x + 2$

Because these two functions are both polynomials, solving this problem is really about collecting like terms and subtracting them. (Just be sure to watch your negative signs!) Here's the math:
$(f - g)(x) = x^2 - 6x + 2 - (2x^2 - 5x) = x^2 - 6x + 2 - 2x^2 + 5x = -x^2 - x + 2.$

19 Find $(f + h)(x)$.

20 Find $(f \cdot g)(x)$.

21 Find $\left(\dfrac{h}{g}\right)(x)$. Does this new function have any undefined values?

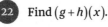 **22** Find $(g+h)(x)$.

Evaluating Composition of Functions

When you are performing the operations of one function on another one, you're creating a *composition of functions*. For instance, if you have two functions, $f(x)$ and $g(x)$, then the composition $f(g(x))$ takes g and puts it in place of x in the formula for $f(x)$. This composition is also written as $(f \circ g)(x)$, and it's basically read right to left; the g function goes into the f function.

For convenience, the same three functions from the preceding section will be used to illustrate how to evaluate the functions that result from composition of functions:

>> $f(x) = x^2 - 6x + 2$

>> $g(x) = 2x^2 - 5x$

>> $h(x) = \sqrt{3x+2}$

Q. Find $f(h(x))$.

EXAMPLE

A. $f(h(x)) = 3x + 4 - 6\sqrt{3x+2}$

Start by substituting the entire h function for every x in the f function:

$f\left(\sqrt{3x+2}\right) = \left(\sqrt{3x+2}\right)^2 - 6\sqrt{3x+2} + 2.$

A square root and a square cancel each other out: $= 3x + 2 - 6\sqrt{3x+2} + 2.$

Simplify by combining any like terms: $= 3x + 4 - 6\sqrt{3x+2}.$

23 Find $(f \circ g)(x)$.

24 Find $(g \circ f)(x)$.

25 Find $h(f(x))$.

26 Find $(f \circ f)(x)$.

27 Find $f(g(-1))$.

28 Find $g(h(3))$.

Working Together: Domain and Range

When you're combining and composing functions (see the two preceding sections) and you want to find out what's happening to the domain and range of the new function, you must keep in mind that in a function, the domain is the input, usually x, and the range is the output, usually y. The truth is that the domain of the given function depends on the operations being performed and the original functions. When combining and composing functions, it's possible that something changes, and it's also possible that nothing does. Typically in pre-calculus, you're asked to find the domain of a combined function and not the range simply because determining the range is often more complicated.

Two of the main points of focus are functions whose domains aren't all real numbers:

>> **Rational functions:** The denominator of any fraction can't be 0, so it's possible that some rational functions are undefined because of this fact. Set the denominator equal to 0 and solve to find the restrictions on your domain.

>> **Square root functions (or any even root):** The *radicand* (what's under the root sign) can't ever be negative, a fact that affects domain. To find out how, set the radicand greater than or equal to 0 and solve. The solution to this inequality is your domain.

Undefined values are also called *excluded values*. When you're asked to find the domain of a combined function, look it over carefully. There isn't just one rule that works all the time for finding a combined function's domain. You just have to take a look at both of the original functions and ask yourself whether their domains have any restrictions. If they do, these restrictions carry through and may be added to those additional restrictions arising from the combined function.

In the following example and problems, you're asked to use those same three functions you've been using for the last two sections to determine domains of some functions resulting from various combinations or compositions of those three functions:

$$f(x) = x^2 - 6x + 2 \qquad g(x) = 2x^2 - 5x \qquad h(x) = \sqrt{3x + 2}$$

Q. Find the domain of $f(h(x))$.

EXAMPLE

A. The domain is all numbers greater than or equal to $-\dfrac{2}{3}$ or $\left[-\dfrac{2}{3}, \infty\right)$.

Take a look at the original two functions first. Because $f(x)$ is a polynomial, there are no restrictions on the domain. However, $h(x)$ is a square root function, which means the radicand has to be positive: $3x + 2 \geq 0$. Solving for x, $3x \geq -2$ and $x \geq -\dfrac{2}{3}$. The new, combined function must honor this domain as well.

29 Find the domain of $(f \circ g)(x)$.

30 Find the domain of $h(f(x))$.

31 Find the domain of $(f + h)(x)$.

32 Find the domain of $\left(\dfrac{h}{g}\right)(x)$.

Unlocking the Inverse of a Function: Turning It Inside Out

An *inverse function* undoes what a function does. You've seen inverse operations before: Addition undoes subtraction, and division undoes multiplication. It shouldn't surprise you, then, that functions have inverses. If $f(x)$ is the original function, then $f^{-1}(x)$ is the notation for the inverse. This notation is special to the inverse function and is *never* meant to represent $\dfrac{1}{f(x)}$.

In the course of your pre-calculus studies, you'll be asked to do three main things with inverses:

» Given a function, graph its inverse.

» Find the inverse of a given function.

» Show that two functions are inverses of each other.

In all three cases, it's all about input and output. If (a, b) is a point in the original function, then (b, a) is a point in the inverse function. Domain and range swap places from a function to its inverse. So if you're asked to graph the inverse function, graph the original and then swap all x and y values in each point to graph the inverse. To find the inverse of a given function, literally take x and y and switch them. After the swap, solve for y. Then change the name of the result to the notation for an inverse function, $f^{-1}(x)$. Lastly, to show that two functions, $f(x)$ and $g(x)$, are inverses of each other, place one inside the other by using composition of functions, $f(g(x))$, and simplify to show that you get x. Then do it the other way around with $g(f(x))$ to make sure it works both ways.

In case that seems a bit confusing, here are a couple example questions so you can see what is meant. You should probably review these before moving on to the practice questions that follow.

EXAMPLE

Q. Find the inverse of $f(x) = 5x - 4$.

A. $f^{-1}(x) = \dfrac{x+4}{5}$.

First, change $f(x)$ to y and switch the x and y: $x = 5y - 4$. Now solve for y: $y = \dfrac{x+4}{5}$. And, finally, change the y to $f^{-1}(x)$: $f^{-1}(x) = \dfrac{x+4}{5}$.

Q. Determine whether $f(x) = 3x - 1$ and $g(x) = \dfrac{x+1}{3}$ are inverses of each other.

A. These two functions are inverses.

First, find $f(g(x))$: $f(g(x)) = 3\left(\dfrac{x+1}{3}\right) = 1$ $= x + 1 - 1 = x$. This satisfies the rule.

Then, moving on to the next one: $g(f(x)) = \dfrac{3x-1+1}{3} = \dfrac{3x}{3} = x$. That one works, too, so these two functions are inverses of each other.

Graph the inverse of $g(x) = \sqrt{x-2}$.

34 Find the inverse of $k(x) = \dfrac{3x}{x-1}$.

 Determine whether $f(x) = x^3 - 1$ and $g(x) = \sqrt[3]{x} + 1$ are inverses of each other.

 Determine whether $f(x) = \dfrac{1-x}{2}$ and $g(x) = 1 - 2x$ are inverses of each other.

Answers to Problems on Functions

Following are the answers to problems dealing with functions. There is also guidance on getting the answers if you need to review the details.

1 Is $f(x) = x^3 - 1$ even, odd, or neither? The answer is neither.

Find $f(-x) = (-x)^3 - 1 = -x - 1$. This isn't the same function as the original one, so the function isn't even. It's also not the exact opposite (sign of each term changed) of the original, so the function isn't odd either. Therefore, this function is neither odd nor even.

2 Determine whether the given graph is even, odd, or neither. The answer is odd.

If you look at the graph upside down, it looks exactly the same — that means it's odd.

3 Sketch half of the graph of $f(x) = \sqrt{x^2 - 4}$ and use symmetry to complete the graph. Graphing the points $(2, 0)$, $\left(3, \sqrt{5}\right) \approx (3, 2.2)$ and $\left(9, \sqrt{77}\right) \approx (9, 8.8)$ helps you determine the curve. See the graph for the answer.

The graph appears to be symmetric and even, but you can check by finding $f(-x) = \sqrt{(-x)^2 - 4} = \sqrt{x^2 - 4}$. The function rule doesn't change at all, which means you have an even function.

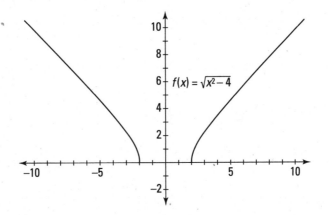

4 Sketch half the graph of $f(x) = 4x^3$ and use symmetry to complete the graph. See the graph for the answer.

You see that $f(-x)$ is $-4x^3$, which is the exact opposite of the original function. Looks like you have an odd graph. Each x gives you a value that's $f(x)$, and each opposite $-x$ gives the opposite $-f(x)$. Plug in some values to get the graph: $f(-3) = -108$, so $f(3) = 108$. $f(2) = 32$, so $f(-2) = -32$. $f(-1) = -4$, so $f(1) = 4$. Put these and as many other points as you want on the graph.

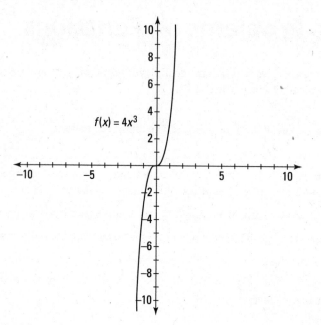

$f(x) = 4x^3$

(5) Graph the function $a(x) = -2(x-1)^2$. See the graph for the answer.

This function takes the parent quadratic graph and moves it to the right by 1. The stretch is 2, making the graph steeper. The negative sign is a reflection, turning the graph upside down. Put all these pieces together to get the graph.

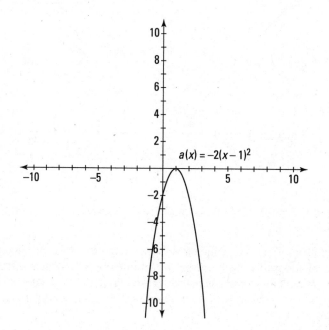

$a(x) = -2(x-1)^2$

6 Graph the function $b(x) = |x + 4| - 1$. See the graph for the answer.

This absolute value function has a horizontal shift of 4 to the left and a vertical shift of 1 down. The coefficient in the front is 1, so this function doesn't go steeper or flatter — the graph has merely been moved.

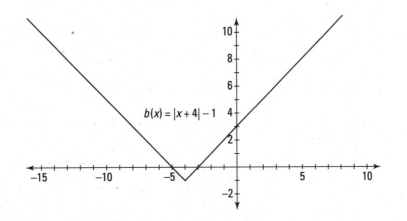

$b(x) = |x + 4| - 1$

7 Graph the function $c(x) = \sqrt{x + 3}$. See the graph for the answer.

This square root function is shifted horizontally to the left by 3.

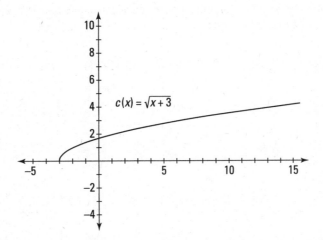

$c(x) = \sqrt{x + 3}$

8 Graph the function $f(x) = -x^2 - 6x$. See the graph for the answer.

This problem actually brings up the topic of conic sections, which is covered in depth in Chapter 12. This function is a polynomial called a parabola. The best way to get this parabola into its standard form, you have to follow a procedure known as *completing the square* (see Chapters 4 and 12 for more information on this procedure). Performing the algebra, the function equation becomes $f(x) = -1(x + 3)^2 + 9$. Now you can use the transformations to determine how the parent curve has changed. You see a horizontal shift of 3 to the left, a vertical shift of 9 upward, and a vertical reflection. Putting all those together, you get what is in the graph.

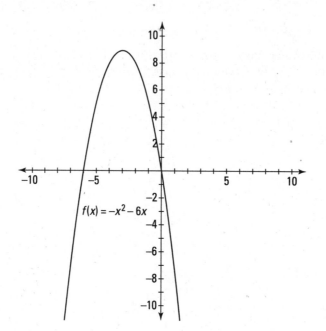

$f(x) = -x^2 - 6x$

9 Graph the function $g(x) - 3$. See the graph for the answer.

To solve this problem, take every single point on the given $g(x)$ function and shift each one down by 3. Actually, if you just use the points where the different segments change direction, you can plot them and then connect those adjacent to one another. You should end up with a graph that looks the one shown.

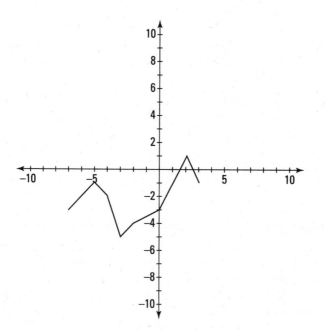

10 Graph the function $2g(x-1)$. See the graph for the answer.

This time, $g(x)$ is shifted to the right by 1 and vertically stretched by a factor of 2. Take the height of each point in the original function and multiply it by 2 to get the new height. For example, the original function passes through the point (2, 4). The height of this point is 4, so when you double that in the new graph, you make the height 8. Do this for the points at the ends of the segments to create the graph.

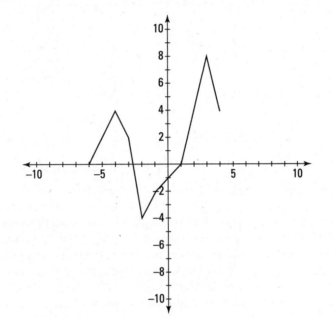

11 Graph the function $q(x) = \dfrac{-2}{(x-4)(x+5)}$. See the graph for the answer.

The vertical asymptotes come from the denominator: $(x+4)(x-5) = 0$. This equation is already neatly factored, so all you have to do is use the zero product property, set each factor equal to 0, and solve. You get $x = 4$ and $x = -5$. Put both of these on the graph as vertical asymptotes. The horizontal asymptote is the x-axis, because the denominator has the greater degree. The intervals you need to take a closer look at are $(-\infty, -5)$, $(-5, 4)$, and $(4, \infty)$.

Pick an x-value from each interval to get an idea of what the graph is doing. For example, when $x = -7$, $y = -0.09$. This point is below the horizontal asymptote. When $x = -4$, $y = 0.25$. This point is above the horizontal asymptote. On the final interval, when $x = 5$, $y = -0.2$; this point is below the horizontal asymptote. Put all the pieces together in the final graph.

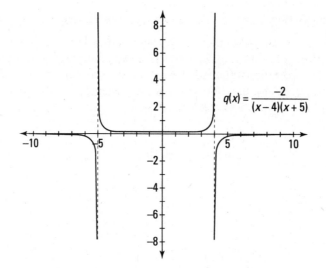

$$q(x) = \frac{-2}{(x-4)(x+5)}$$

(12) Graph the function $r(x) = \dfrac{x+3}{x^2 - x - 6}$. See the graph for the answer.

First up are vertical asymptotes. Set $x^2 - x - 6 = 0$ and factor to get $(x-3)(x+2) = 0$. Set each factor equal to 0 and solve. You get $x = 3$ and $x = -2$. Add these two vertical asymptotes to your graph. Next up is the horizontal asymptote. Because the denominator has the greater degree, the horizontal asymptote is the x-axis again. Notice, also, that there is a variable in the numerator, so there may be an x-intercept. Set the numerator equal to 0 and solve. $x + 3 = 0$ tells you that $x = -3$ and you have an intercept at $(-3, 0)$. The graph crosses the x-axis. Use this fact to set up the intervals: $(-\infty, -3)$, $(-3, -2)$, $(-2, 3)$, and $(3, \infty)$. The points in the intervals are, respectively, below, above, below, and above the horizontal asymptote.

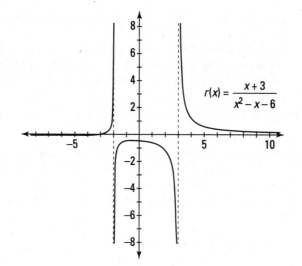

$$r(x) = \frac{x+3}{x^2 - x - 6}$$

13 Graph the function $t(x) = \dfrac{x^2 - 5x}{x^2 - 4x - 21}$. See the graph for the answer.

Find the vertical asymptotes for this one by factoring the denominator. If $x^2 - 4x - 21 = 0$, then $(x-7)(x+3) = 0$. This gives you two solutions or vertical asymptotes: $x = 7$ and $x = -3$. The degrees in the numerator and denominator are the same, so the horizontal asymptote is $y = 1$. Put the asymptotes on the graph and then pick x-values and x-intercepts to get the graph.

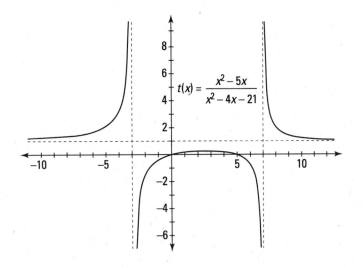

14 Graph the function $u(x) = \dfrac{x^2 - 10x - 24}{x + 1}$. See the graph for the answer.

This is an interesting problem because there's no horizontal asymptote, but you have an oblique asymptote — the numerator has a degree one greater than the denominator. Use long division to divide the numerator by the denominator. You get the quotient $x - 11$ (by ignoring the remainder); graph this as an equation, $y = x - 11$, with a dotted line to mark your oblique asymptote. Next, add the vertical asymptote by solving the equation $x + 1 = 0$ to get $x = -1$. Finally, try some values on each interval to get the graph. In particular, you have two x-intercepts and a y-intercept. They're at (12, 0), (−2, 0) and (0, −24), respectively.

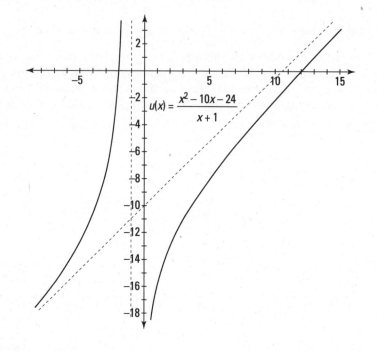

$$u(x) = \frac{x^2 - 10x - 24}{x + 1}$$

15 Graph $g(x) = \begin{cases} \sqrt{x+3} & \text{if } x \le -1 \\ (x+3)^2 & \text{if } x > -1 \end{cases}$. See the graph for the answer.

Take a look at each interval of the domain to determine the graph's shape. For this piecewise function, the first rule only applies when $x \le -1$. So this part of the graph looks like a square root graph shifted 3 to the left. The bottom rule is used when $x > -1$, which makes this part of the graph a parabola that's shifted 3 to the left.

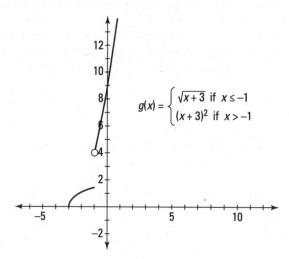

$$g(x) = \begin{cases} \sqrt{x+3} & \text{if } x \le -1 \\ (x+3)^2 & \text{if } x > -1 \end{cases}$$

Sometimes it helps to lightly sketch in the whole graph determined by a rule and then erase the part you don't need.

TIP

16 Graph $h(x) = \begin{cases} \frac{1}{2}x - 4 & \text{if } x \le -2 \\ 3x + 3 & \text{if } -2 < x < 2 \\ 4 - x & \text{if } x \ge 2 \end{cases}$ See the graph for the answer.

The top piece is a linear function that's defined only when $x \le -2$. The second rule is also a linear function, defined between -2 and 2. The bottom rule is yet another linear function, defined when $x \ge 2$.

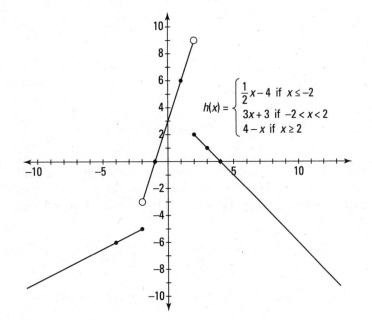

$h(x) = \begin{cases} \frac{1}{2}x - 4 & \text{if } x \le -2 \\ 3x + 3 & \text{if } -2 < x < 2 \\ 4 - x & \text{if } x \ge 2 \end{cases}$

17 Graph $m(x) = \begin{cases} x^3 + 2 & \text{if } x < 0 \\ x^2 + 2 & \text{if } 0 \le x < 2 \\ x + 2 & \text{if } x \ge 2 \end{cases}$ See the graph for the answer.

The first rule is a cubic function that's shifted up by 2 — its right endpoint should be open. However, when you graph the second rule, you see that it's a parabola that's shifted up by 2. Its left endpoint overlaps the right endpoint of the first part of the graph, filling the hole that was there, and the graph carries on until $x = 2$, where it gets broken again. The third rule follows the linear function to the right of $x = 2$.

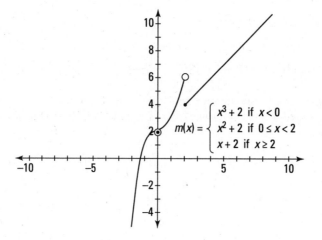

$$m(x) = \begin{cases} x^3 + 2 & \text{if } x < 0 \\ x^2 + 2 & \text{if } 0 \le x < 2 \\ x + 2 & \text{if } x \ge 2 \end{cases}$$

18) Graph $n(x) = \begin{cases} |x-1| & \text{if } x < -3 \\ -3 & \text{if } x = -3 \\ |x|-1 & \text{if } x > -3 \end{cases}$. See the graph for the answer.

This piecewise function is different because the middle part is defined at just one point. When $x = 3$, $y = -3$. That's it. The first rule follows that for the absolute value graph that has been shifted to the right by 1. The third rule is also an absolute value graph, but it has been shifted down 1.

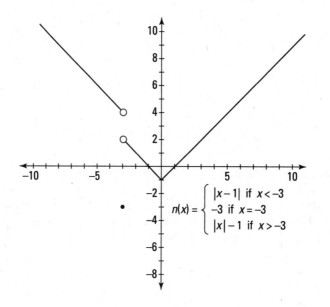

$$n(x) = \begin{cases} |x-1| & \text{if } x < -3 \\ -3 & \text{if } x = -3 \\ |x|-1 & \text{if } x > -3 \end{cases}$$

19) Find $(f+h)(x)$. The answer is $(f+h)(x) = x^2 - 6x + 2 + \sqrt{3x+2}$.

Take the f function and add the h function to it. Because one is a polynomial and the other is a square root, there are no like terms. The answer is therefore $(f+h)(x) = x^2 - 6x + 2 + \sqrt{3x+2}$.

(20) Find $(f \cdot g)(x)$. The answer is $(f \cdot g)(x) = 2x^4 - 17x^3 + 34x^2 - 10x$.

Start off by writing out what you've been asked to find — the product of f and g: $(f \cdot g)(x) = (x^2 - 6x + 2)(2x^2 - 5x)$. Distribute each term of the left polynomial to each term of the right polynomial to get $2x^4 - 5x^3 - 12x^3 + 30x^2 + 4x^2 - 10x$. Next, combine the like terms $(f \cdot g)(x) = 2x^4 - 17x^3 + 34x^2 - 10x$.

(21) Find $\left(\dfrac{h}{g}\right)(x)$. Does this new function have any undefined values? The answer is

$\left(\dfrac{h}{g}\right)(x) = \dfrac{\sqrt{3x+2}}{2x^2 - 5x}$, and yes, there are undefined values — specifically, $x = 0$ and $x = \dfrac{5}{2}$ which create a 0 in the denominator. It is also undefined when $x < -\dfrac{2}{3}$, because those x-values create negative numbers under the radical.

You're being asked to find the quotient of h and g, with h on the top and g on the bottom.

They're different types of functions, and you can't simplify the fraction, which means your answer is, simply, $\left(\dfrac{h}{g}\right)(x) = \dfrac{\sqrt{3x+2}}{2x^2 - 5x}$.

To find the undefined values, set the denominator equal to 0 to start: $2x^2 - 5x = 0$. Solve by factoring out the greatest common factor: $x(2x - 5)$. This has two solutions, $x = 0$ and $x = \dfrac{5}{2}$. These are two of the undefined values. To determine the numbers creating a negative under the radical, just solve the inequality $3x + 2 < 0$.

(22) Find $(g + h)(x)$. The answer is $(g + h)(x) = 2x^2 - 5x + \sqrt{3x+2}$.

No like terms exist because you're adding a polynomial and a square root. That means the answer is $(g + h)(x) = 2x^2 - 5x + \sqrt{3x+2}$.

(23) Find $(f \circ g)(x)$. The answer is $(f \circ g)(x) = 4x^4 - 20x^3 + 13x^2 + 30x + 2$.

Take the g function and replace every x in the function f with its rule: $(2x^2 - 5x)^2 - 6(2x^2 - 5x) + 2$. Multiply everything out, and then combine like terms: $4x^4 - 20x^3 + 25x^2 - 12x^2 + 30x + 2 = 4x^4 - 20x^3 + 13x^2 + 30x + 2$.

(24) Find $(g \circ f)(x)$. The answer is $(g \circ f)(x) = 2x^4 - 24x^3 + 75x^2 - 18x - 2$.

This time, place f into function g wherever it says x: $(g \circ f)(x) = 2(x^2 - 6x + 2)^2 - 5(x^2 - 6x + 2)$. Square the first trinomial, and then distribute the 2. Distribute the –5 through the right trinomial. Then combine all the like terms from both processes.

$$2(x^2 - 6x + 2)^2 - 5(x^2 - 6x + 2)$$
$$2(x^4 - 6x^3 + 2x^2 - 6x^3 + 36x^2 - 12x + 2x^2 - 12x + 4) - 5(x^2 - 6x + 2)$$
$$= 2(x^4 - 12x^3 + 40x^2 - 24x + 4) - 5(x^2 - 6x + 2)$$
$$= 2x^4 - 24x^3 + 80x^2 - 48x + 8 - 5x^2 + 30x - 10$$
$$= 2x^4 - 24x^3 + 75x^5 - 18x - 2$$

(25) Find $h(f(x))$. The answer is $h(f(x)) = \sqrt{3x^2 - 18x + 8}$.

Substitute the rule for f for x in the h function: $h(f(x)) = \sqrt{3(x^2 - 6x + 2) + 2}$. Distribute the 3, and then add the 2: $\sqrt{3x^2 - 18x + 6 + 2} = \sqrt{3x^2 - 18x + 8}$.

(26) Find $(f \circ f)(x)$. The answer is $(f \circ f)(x) = x^4 - 12x^3 + 34x^2 + 12x - 6$.

Substitute the rule for f into itself everywhere it says x: $(x^2 - 6x + 2)^2 - 6(x^2 - 6x + 2) + 2$.
Square the trinomial, and distribute the -6: $x^4 - 12x^3 + 40x^2 - 24x + 4 - 6x^2 + 36x - 12 + 2$.
Combine like terms to end up with this answer: $(f \circ f)(x) = x^4 - 12x^3 + 34x^2 + 12x - 6$.

(27) Find $f(g(-1))$. The answer is 9.

First find $g(-1)$, and then find f of the result. $g(-1) = 2(-1)^2 - 5(-1) = 2 \cdot 1 + 5 = 7$. Then, $f(7) = (7)^2 - 6(7) + 2 = 49 - 42 + 2 = 9$.

(28) Find $g(h(3))$. The answer is $22 - 5\sqrt{11}$.

Start with $h(3) = \sqrt{3(3) + 2} = \sqrt{11}$. Now plug this value into g: $g(\sqrt{11}) = 2(\sqrt{11})^2 - 5(\sqrt{11}) = 2(11) - 5\sqrt{11} = 22 - 5\sqrt{11}$.

(29) Find the domain of $(f \circ g)(x)$. The domain is all real numbers.

Question 23 has you find the composition $(f \circ g)(x)$. The result is $(f \circ g)(x) = 4x^4 - 20x^3 + 13x^2 + 30x + 2$, which is a polynomial. Its domain is all real numbers.

(30) Find the domain of $h(f(x))$. The domain is $x \leq \dfrac{9 - \sqrt{57}}{3}$ or $x \geq \dfrac{9 + \sqrt{57}}{3}$.

In Question 25, you find that $h(f(x)) = \sqrt{3x^2 - 18x + 8}$, which has a quadratic expression under the square root. The quantity under the radical must be nonnegative. Find where $3x^2 - 18x + 8$ is nonnegative by setting it greater than or equal to zero and solving the resulting quadratic inequality by using the quadratic formula. Using the quadratic formula on $3x^2 - 18x + 8 = 0$ your solutions are $x = \dfrac{9 \pm \sqrt{57}}{3}$. The values of x are approximately 5.5 and 0.5. You need to test values less than 0.5, between 0.5 and 5.5, and values greater than 5.5 to see if they result in nonnegative results – which is needed under the radical. When x is less than 0.5 or greater than 5.5, the quadratic is positive, so the domain consists of $x \leq \dfrac{9 - \sqrt{57}}{3} \approx 0.5$ or $x \geq \dfrac{9 + \sqrt{57}}{3} \approx 5.5$.

(31) Find the domain of $(f + h)(x)$. The domain is $x \geq -\dfrac{2}{3}$.

$(f + h)(x) = x^2 - 6x + 2 + \sqrt{3x + 2}$. The domain contains all x resulting in a positive number or 0 under the radical. $3x + 2 \geq 0$ when $x \geq -\dfrac{2}{3}$.

(32) Find the domain of $\left(\dfrac{h}{g}\right)(x)$. The domain is $x \geq -\dfrac{2}{3}$, except $x = 0$ and $x = \dfrac{5}{2}$.

$\left(\dfrac{h}{g}\right)(x) = \dfrac{\sqrt{3x + 2}}{2x^2 - 5x}$. The square root in the numerator restricts the domain to $x \geq -\dfrac{2}{3}$. The polynomial in the denominator has undefined values: $x = 0$ and $x = \dfrac{5}{2}$. These are both in the restricted domain, so they become part of the answer. In interval notation, the domain is written that x is: $\left[-\dfrac{2}{3}, 0\right) \cup \left(0, \dfrac{5}{2}\right) \cup \left(\dfrac{5}{2}, \infty\right)$.

(33) Graph the inverse of $g(x) = \sqrt{x-2}$. See the graph for the answer.

Using the graph of $g(x) = \sqrt{x-2}$, which is a square root function shifted 2 to the right, you find the points (2, 0), (3, 1), and (6, 2) on that graph. The inverse is the reflection of the original function across the line $y = x$. Points on the inverse are (0, 2), (1, 3), and (2, 6) — all with the x and y coordinates reversed. The inverse function is $g^{-1}(x) = x^2 + 2$, but only for values of x greater than or equal to 0.

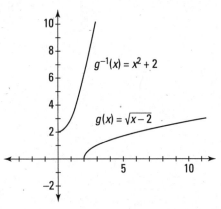

(34) Find the inverse of $k(x) = \dfrac{3x}{x-1}$. The answer is $k^{-1}(x) = \dfrac{x}{x-3}$.

Replace $k(x)$ with y. Then change the y to x and the x's in the function rule to y's. Solve for y by first multiplying both sides of the equation by the denominator of the fraction:

$x(y-1) = \dfrac{3y}{y-1} \cdot (y-1) \;\rightarrow\; xy - x = 3y$. Now subtract 3$y$ from each side and add x to each side,

and you get: $xy - 3y = x$. Factor the left side, and then divide each side by $(x-3)$:

$(x-3)y = x \;\rightarrow\; \dfrac{(x-3)y}{x-3} = \dfrac{x}{x-3}$. The resulting $y = \dfrac{x}{x-3}$ is rewritten replacing the y with $k^{-1}(x)$.

(35) Determine whether $f(x) = x^3 - 1$ and $g(x) = \sqrt[3]{x} + 1$ are inverses of each other. The answer is that they're not inverses.

Testing the compositions, you start with: $(f \circ g)(x) = \left(\sqrt[3]{x} + 1\right)^3 - 1 = x + 3x^{2/3} + 3x^{1/3} + 1 - 1 = x + 3x^{2/3} + 3x^{1/3}$. This doesn't simplify to be x, so you can stop. They're not inverses.

(36) Determine whether $f(x) = \dfrac{1-x}{2}$ and $g(x) = 1 - 2x$ are inverses of each other. The answer is yes, they are inverses.

Testing the compositions: $(f \circ g)(x) = \dfrac{1-(1-2x)}{2} = \dfrac{2x}{2} = x$ and $(g \circ f)(x) = 1 - 2\left(\dfrac{1-x}{2}\right)$

$= 1 - (1-x) = x$. They both result in x. They're inverses.

Chapter **4**

Searching for Roots

A *polynomial* is an algebraic expression of one or more terms involving a sum of powers of a variable multiplied by coefficients. The powers are always nonnegative integers. The highest exponent on any term in a polynomial is its *degree*. In this chapter, you'll find practice on solving polynomial equations to find the solutions, which are also called *roots* or *zeros*. The chapter starts with a review of solving *quadratic equations* — polynomials where the highest exponent is two. Then you move into polynomial equations with higher degrees and you see how to solve them. Finally, you take a look at using roots to factor polynomials and then graph them.

Factoring a Factorable Quadratic

Before getting started on the nitty-gritty, here's some vocabulary you should know for work in this chapter (and hereafter):

» **Standard form:** What most use to write an equation: $ax^2 + bx + c = 0$

» **Quadratic term:** The term with the second degree: ax^2

» **Linear term:** The term with the linear degree: bx

» **Constant:** The term with zero degree: c

» **Leading coefficient:** The number multiplying the term with the highest degree: a

REMEMBER

In math, the process of breaking down a polynomial into the product of two or more polynomials with a smaller degree is called *factoring*. In general, factoring is often easy for quadratic equations and is the first thing you want to try when asked to solve second-degree polynomial equations. Some types of factoring (like the difference of cubes or grouping — more on those later in this chapter) may work on higher-degree polynomials, and you should always check to see whether this is the case first. When presented with a polynomial equation and asked to solve it, you should always try the following methods of factoring, in order:

>> **Finding the greatest common factor:** The *greatest common factor*, or GCF, is the biggest expression that will divide into all the other terms. It's a little like doing the distributive property backwards.

   Look at all those factors to see what they share in common (that's your GCF), factor the GCF out from every term. Write the result by putting the GCF outside a set of parentheses, and leave the terms that aren't the GCF inside the parentheses.

>> **Working with a binomial polynomial:** If the polynomial has two terms, check to see whether it's a difference of squares or the sum or difference of cubes. Just the first of these is quadratic.

   • **Difference of squares:** $a^2 - b^2 = (a-b)(a+b)$

   • **Difference of cubes:** $a^3 - b^3 = (a-b)\left(a^2 + ab + b^2\right)$

   • **Sum of cubes:** $a^3 + b^3 = (a+b)\left(a^2 - ab + b^2\right)$

>> **Recognizing a perfect square trinomial polynomial:** When the same two binomials are multiplied together, you have a recognizable pattern:

   $$a^2 \pm 2ab + b^2 = (a \pm b)^2$$

>> **Working with a trinomial polynomial:** A trinomial in the form $ax^2 + bx + c$ may or may not be factorable. Sometimes the factorization is clear. Other times you have to try several combinations. A nice method to use is the *box method*. Follow these steps to use this method:

   1. **Draw a two-by-two box; put the first term in the upper left and the last term in the lower right.**

   2. **Find the product of the first and last terms.**

   3. **Find a pair of factors of the product whose sum is the middle term.**

      Write those factors in the other two boxes.

   4. **Find the GCF of each row and column, and write it to the left and below the box.**

      These are the terms that go in the factorization of the trinomial.

Once you factor the polynomial, you can use the zero product property to solve it by setting each factor equal to 0 and solving.

EXAMPLE

Q. Solve the equation $2x^2 - 5x - 12 = 0$.

A. $x = -\dfrac{3}{2}, x = 4$. Draw a two-by-two box, and put $2x^2$ and -12 in their places.

$2x^2$	
	-12

The product of $2x^2$ and -12 is $-24x^2$. The two factors of $-24x^2$ that add up to the middle term are $-8x$ and $3x$. Put them in the empty boxes (doesn't matter which goes where).

$2x^2$	$-8x$
$3x$	-12

Write the GCF of each row and column to the left and below the box.

$2x^2$	$-8x$	$2x$
$3x$	-12	3
x	-4	

Write the factorization using the outside values: $2x^2 - 5x - 12 = (2x + 3)(x - 4)$.

Set the factors equal to 0 to solve for x: $2x + 3 = 0$ yields $x = -\dfrac{3}{2}$ and $x - 4 = 0$ gives you $x = 4$.

Q. Solve the equation $3x^2 - 48 = 0$.

A. $x = 4$, $x = -4$. Always check for the GCF first and factor it out: $3(x^2 - 16) = 0$. Now recognize the binomial as a difference of squares that factors again: $3(x - 4)(x + 4) = 0$. Set each factor equal to 0 and solve: $x - 4 = 0$ gives you $x = 4$, and $x + 4 = 0$ yields $x = -4$.

---

1 Solve the equation $6y^2 + 13y = 5$.

2 Solve the equation $16m^2 - 8m + 1 = 0$.

3 Solve the equation $48x^2 - 10x - 3 = 0$.

4 Solve the equation $\frac{1}{6}x^2 + \frac{2}{3}x = 2$.

Solving a Quadratic Polynomial Equation

What happens when a quadratic equation doesn't factor? You're done, right? Well, not quite. You have two more methods you can use: the quadratic formula, and completing the square. When you graph quadratics, as you do in Chapters 3 and 12, the easiest method to use is to complete the square and then use the rules of transforming a parent function to get the graph.

Completing the square

Here are the steps for completing the square:

1. Make sure the quadratic is written in standard form: $ax^2 + bx + c = 0$

2. Add (or subtract) the constant term from both sides: $ax^2 + bx = -c$

3. Factor out the leading coefficient from the quadratic term and the linear term:

    $$a\left(x^2 + \frac{b}{a}x\right) = -c$$

4. Divide each side of the equation by the leading coefficient: $\dfrac{a\left(x^2 + \frac{b}{a}x\right)}{a} = \dfrac{-c}{a}$ simplifies to $x^2 + \dfrac{b}{a}x = -\dfrac{c}{a}$

5. Divide the new linear coefficient by two: $\dfrac{b/a}{2} = \dfrac{b}{2a}$; square the result: $\left(\dfrac{b}{2a}\right)^2 = \dfrac{b^2}{4a^2}$.

6. Add the squared term to each side of the equation: $x^2 + \dfrac{b}{a}x + \dfrac{b^2}{4a^2} = -\dfrac{c}{a} + \dfrac{b^2}{4a^2}$

7. Factor the trinomial on the left side and add the terms on the right side:

    $$\left(x + \frac{b}{2a}\right)^2 = -\frac{4ac}{4a^2} + \frac{b^2}{4a^2} = \frac{b^2 - 4ac}{4a^2}$$

8. Take the square root of both sides: $\sqrt{\left(x+\dfrac{b}{2a}\right)^2}=\pm\sqrt{\dfrac{b^2-4ac}{4a^2}}$. Simplify:

$$x+\frac{b}{2a}=\frac{\pm\sqrt{b^2-4ac}}{2a}$$

9. Solve for x: $x=-\dfrac{b}{2a}\pm\dfrac{\sqrt{b^2-4ac}}{2a}$

Quadratic formula

Of course, if you're familiar with the quadratic formula you will recognize the preceding steps — they're the derivation of the quadratic formula. All you have to do is add the fractions and end up with the quadratic formula:

$$x=\frac{-b\pm\sqrt{b^2-4ac}}{2a}$$

EXAMPLE

Q. Solve the equation $5x^2-12x-2=0$.

A. $x=\dfrac{6\pm\sqrt{46}}{5}$. This equation doesn't factor, so you use the quadratic formula to solve it.

$a=5$, $b=-12$, $c=-2$. Plug these values into the quadratic formula:

$$x=\frac{-b\pm\sqrt{b^2-4ac}}{2a}=\frac{-(-12)\pm\sqrt{(-12)^2-4(5)(-2)}}{2(5)}$$

Simplifying, $=\dfrac{12\pm\sqrt{144+40}}{10}=\dfrac{12\pm\sqrt{184}}{10}=\dfrac{12\pm\sqrt{4\cdot46}}{10}=\dfrac{12\pm2\sqrt{46}}{10}=\dfrac{6\pm\sqrt{46}}{5}$.

---

5   Solve $x^2-10=2x$.

6   Solve $7x^2-x+2=0$.

7 Solve $x^2 - 4x - 7 = 0$ by completing the square.

8 Solve $-2.31x^2 - 4.2x + 6.7 = 0$.

Solving High-Order Polynomials

The greater the degree of your given polynomial, the more challenging it is to solve the equation by factoring. You should always still try that first because you never know . . . it may actually work! When factoring fails, however, you begin anew with another process for finding the roots. You find several options in this section.

REMEMBER Always begin by finding the degree of the polynomial because it gives you some very important information about your graph. The degree of the polynomial tells you the maximum number of roots — it's that easy. A fourth-degree polynomial will have up to, but no more than, four roots.

Factoring by grouping

When working with a polynomial that has more than three terms, try grouping the polynomial into two sets of two. Find the GCF for each set and factor it out. Find the GCF of the two remaining expressions and factor it out. You end up with the product of two binomials, exactly what you want! For example, in the polynomial equation $2x^3 + 7x^2 - 8x - 28 = 0$ the first two terms have a common factor of x^2, and the last two have a common factor of -4. Factoring, you have $x^2(2x+7) - 4(2x+7) = 0$. The two new terms have a common factor of $(2x+7)$, so you factor again to get $(2x+7)(x^2-4) = 0$. The difference of squares can now also be factored, and the roots determined rather easily.

Determining positive and negative roots: Descartes' Rule of Signs

When you have a polynomial with real coefficients, you can use *Descartes' Rule of Signs* to determine the possible number of positive and negative real roots. All you have to do is count! First, make sure that the polynomial $f(x)$ has real coefficients and the terms written in descending order, from highest to lowest degree. Look at the sign of each term and count how many times the sign changes from positive to negative and vice versa as you view the polynomial from left to right. The number of sign changes represents the maximum number of positive real roots. The rule also says that the possible number of positive real roots decreases by 2 over and over again until you end up with 1 or 0 (more on this in the next section). This gives you the list of the possible number of real positive roots. For example, the function $f(x) = 5x^5 - 4x^4 + 3x^3 + 2x^2 - x - 11$ has 3 or 1 positive real root.

Descartes also figured out that if you take a look at $f(-x)$ and count again, you discover the maximum number of negative real roots. So, since $f(-x) = -5x^5 - 4x^4 - 3x^3 + 2x^2 + x - 11$, you find two sign changes, so there are either 2 or 0 negative real roots.

Counting on imaginary roots

Complex roots happen in a quadratic equation that has real coefficients when the *discriminant* (the part of the quadratic formula under the root sign) is negative. The $\pm$ in the quadratic formula also tells you that there are two of these roots, always in pairs. This is why you subtract by 2 in the preceding section; you have to account for the fact that some of the roots may come in pairs of complex numbers. In fact, the complex pairs will always be *complex conjugates* of each other — if one root is $a + bi$, the other one is $a - bi$. For example, looking at the polynomial $g(x) = x^4 - 5x^3 + 8x^2 - 10x - 12$, you see that there are three or one possible positive real roots. Since it factors: $g(x) = (x - 6)(x + 1)(x^2 + 2)$, you can see that there's just one positive real root and the complex conjugate $0 \pm i\sqrt{2}$.

Getting the rational roots

When all the coefficients of a polynomial are integers, the *Rational Root Theorem* helps you narrow down the possibilities for the roots even further. Right now, if you've gone through all the steps, you know only the total number of possible roots, how many might be positive or negative real, and how many might be complex. That still leaves an infinite number of possibilities for the values of the roots! The Rational Root Theorem helps you because it finds the possible roots that are *rational* (those that can be written as fractions). The problem with the theorem? Not all roots are rational. Keep in mind that some (or all) of the roots may be irrational or complex numbers.

To use the Rational Root Theorem, take all the factors of the constant term and divide by all the factors of the leading coefficient; both positive and negative divisors must be included. This produces a list of fractions that are all possibilities for roots. For example, in the polynomial $h(x) = 2x^4 - 3x^3 + x - 5$, the list of possible rational roots is: $\pm 1, \pm 5, \pm \dfrac{1}{2}, \pm \dfrac{5}{2}$.

Finding roots through synthetic division

Armed with a list of possible rational roots, you can pick one fraction and try to find its roots through the process of synthetic division, as described in this section. If this tactic works, the quotient formed is a *depressed polynomial.* No, it's not sad — but its degree will be less than the one you started with. You use this quotient to find the next one, each time lessening the degree, which narrows down the roots you have to find. At some point, your polynomial will end up as a quadratic equation, which you can solve using factoring or the quadratic formula. Now that's clever! If the root you try works, you should always try it again to see if it's a root with *multiplicity* — that is, a root that's used more than once.

Here are the steps to use for synthetic division:

1. **Write the polynomial in descending order. If any degrees are missing, fill in the gaps with zeros.**

2. **Write the number of the root you're testing outside the synthetic division sign. Write the coefficients of the polynomial in descending order and include any zeros from Step 1 inside the synthetic division sign.**

3. **Drop the first coefficient down.**

4. **Multiply the root on the outside and this coefficient. Write this product above the synthetic division line.**

5. **Add the next coefficient and the product from Step 4. This answer goes below the line.**

6. **Multiply the root on the outside and the answer from Step 5.**

7. **Repeat over and over again until you use all the coefficients.**

This process is easier to see with an example, like the one that follows. Just know that when you do synthetic division, you end up with a list of roots that actually work in the polynomial.

Q. Find the roots of the equation $x^3 + x^2 - 5x + 3 = 0$.

EXAMPLE **A.** $x = 1$ (double root), $x = -3$. The different processes described in this section are used for this example question.

The number of roots: First, this equation is third degree, so it may have up to three different roots.

Descartes' Rule of Signs: Next, by looking at $f(x) = x^3 + x^2 - 5x + 3 = 0$, you notice that the sign changes twice. This means there could be two or zero positive real roots. Next, look at $f(-x) = -x^3 + x^2 + 5x + 3 = 0$ and notice the sign changes only once, giving you only one negative real root.

Complex roots: So if two roots are positive and one is negative, that leaves none left over that are complex. But if none are positive and one is negative, that leaves two complex roots.

Rational Root Theorem: Take all the factors of 3 (the constant term) and divide by all the factors of 1 (the leading coefficient) to determine the possible rational roots: $\pm 1, \pm 3$.

Synthetic division: Pick a root, any root, and use synthetic division to test and see whether it actually is a root. A good choice to start with is $x = 1$:

$$
\begin{array}{r|rrrr}
1 & 1 & 1 & -5 & 3 \\
  &   & 1 & 2 & -3 \\
\hline
  & 1 & 2 & -3 & 0
\end{array}
$$

The last column on the right is the remainder; because it's 0, you know you have one root: $x = 1$. Also notice that the other numbers are the coefficients of the depressed polynomial you're now working with: $x^2 + 2x - 3 = 0$. Because this is a quadratic, it's recommended shifting gears and factoring it to $(x+3)(x-1) = 0$ to be able to use the zero product property to solve and get $x = -3$ and $x = 1$ (again — making it a double root!).

Q. Solve the equation $x^3 + 8x^2 + 22x + 20 = 0$.

A. $x = -2, x = -3, x = -3 + i, x = -3 - i$. This equation is a third degree, so it will have a maximum of three roots. Looking at $f(x) = x^3 + 8x^2 + 22x + 20 = 0$ you see no changes in sign, so none of the roots are positive. Looking at $f(-x) = -x^3 + 8x^2 - 22x + 20 = 0$ reveals that either one or three of them are negative. The Rational Root Theorem generates this list of fractions (you only need to consider negatives): $-1, -2, -4, -5, -10, -20$. Start off with $x = -2$ to discover one of your roots:

$$
\begin{array}{r|rrrr}
-2 & 1 & 8 & 22 & 20 \\
   &   & -2 & -12 & -20 \\
\hline
   & 1 & 6 & 10 & 0
\end{array}
$$

The reduced polynomial you're now working with is $x^2 + 6x + 10 = 0$. This quadratic doesn't factor, so you use the quadratic formula to find that the last two roots are indeed complex: $x = -3 \pm i$.

---

 9 Solve the equation $2x^3 + 3x^2 - 18x + 8 = 0$.

10 Solve the equation $4x^4 + x^3 - 108x - 27 = 0$.

11 Solve the equation $x^3 + 7x^2 + 13x + 4 = 0$.

12 Find the roots of the polynomial
$x^4 + 10x^3 + 38x^2 + 66x + 45 = 0$.

Using Roots to Create an Equation

The *factor theorem* says that if you know the root of a polynomial, then you also know a factor of the polynomial. These two go back and forth, one to the other — roots and linear factors are interchangeable. When asked to factor a polynomial with a degree higher than two, you can use the techniques described in the earlier sections to find the roots and then write the factors using the roots found.

TIP

If $x = c$ is a root, then $x - c$ is a factor and vice versa. And if $x = -c$ is a root, then $x - (-c) = x + c$ is a factor.

EXAMPLE

Q. Use the roots of $x^3 + x^2 - 5x + 3 = 0$ to factor the equation.

A. $(x-1)^2 (x+3) = 0$. This is from an example in the last section. You found that the roots are $x = 1$ (double root) and $x = -3$. Using the factor theorem,

if $x = 1$ is a root, then $x - 1$ is a factor (twice); and if $x = -3$ is a root, then $x + 3$ is a factor. This means that $x^3 + x^2 - 5x + 3 = 0$ factors to $(x-1)^2 (x+3) = 0$.

Find the lowest order polynomial with leading coefficient as 1 that has $-3, -2, 4$, and 6 as its roots.

 Find the lowest order polynomial with leading coefficient as 1 that has $2, 4 + 3i$, and $4 - 3i$ as its roots.

 Factor the polynomial $6x^4 - 7x^3 - 18x^2 + 13x + 6$.

 Factor the polynomial $x^4 + 10x^3 + 38x^2 + 66x + 45$.

Graphing Polynomials

Once you have a list of the roots of your polynomial, you've done the hardest part of graphing the polynomial. Remember that real roots or zeros are x-intercepts — you now know where the graph crosses the x-axis. Follow these steps to get to the graph:

1. **Mark the x-intercepts on your graph.**

2. **Find the y-intercept by letting $x = 0$. The shortcut? It will always be the constant term.**

3. **Use the leading coefficient test to determine which of the four possible ways the ends of your graph will point:**

- If the degree of the polynomial is even and the leading coefficient is positive, both ends of the graph will point up.

- If the degree of the polynomial is even and the leading coefficient is negative, both ends of the graph will point down.

- If the degree of the polynomial is odd and the leading coefficient is positive, the left side of the graph will point down and the right side will point up.

- If the degree of the polynomial is odd and the leading coefficient is negative, the left side of the graph will point up and the right side will point down.

4. **Figure out what happens in between the x-intercepts by picking any x-value on each interval and plugging it into the function to determine whether it's positive (and, therefore, above the x-axis) or negative (below the x-axis).**

5. **Plot the graph by using all the information you've determined.**

Q. Graph the equation $f(x) = x^3 + x^2 - 5x + 3$.

EXAMPLE **A.** See the graph. This is the example from earlier sections. You found that the roots are $x = 1$ (double root) and $x = -3$. The y-intercept is the constant, or $(0, 3)$. The leading coefficient test tells you the graph starts by pointing down and ends by pointing up. The double root at $x = 1$ makes the graph "bounce" and not cross there.

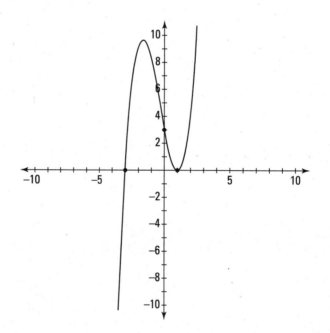

Q. Graph the equation $f(x) = x^3 + 8x^2 + 22x + 20$.

A. See the graph. This is another earlier example. You found one real root of $x = -2$, as well as the complex conjugates $x = -3 \pm i$. The leading coefficient test tells you the graph starts by pointing down and ends by pointing up. There's the one x-intercept (-2, 0) and the y-intercept (0, 20).

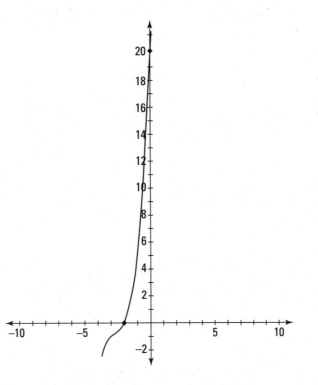

17 Graph $f(x) = x^4 + 2x^3 - 13x^2 - 14x + 24$.

18 Graph $f(x) = 6x^4 - 7x^3 - 18x^2 + 13x + 6$.

Graph $f(x) = 12x^4 + 13x^3 - 20x^2 + 4x$.

20 Graph $f(x) = x^4 + 10x^3 + 38x^2 + 66x + 45$.

Answers to Problems on Roots and Degrees

Following are the answers to questions presented in this chapter. They each also have explanations for the problems.

(1) Solve the equation $6y^2 + 13y = 5$. The answer is $y = \frac{1}{3}, -\frac{5}{2}$.

Begin with any quadratic equation by getting 0 on one side of the equation. In this case, subtract 12 from both sides: $6y^2 + 13y - 5 = 0$. Using the box method, begin by putting the first and last terms in the top left and bottom right corners. The product of the first and last terms is $-30y^2$. Find two factors of $-30y^2$ that add up to the middle term, $13y$. That would be $15y$ and $-2y$. Place those terms in the empty squares. Write the GCF of each row and column to the right and below the box. These are the terms in the factorization.

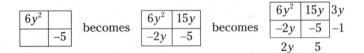

$6y^2 + 13y - 5 = (3y - 1)(2y + 5) = 0$. Setting the factors equal to 0, you get $y = \frac{1}{3}, -\frac{5}{2}$.

(2) Solve the equation $16m^2 - 8m + 1 = 0$. The answer is $m = \frac{1}{4}$.

Why is there only one answer? Oh right, it's a double root, probably. Factor it to find out. This is a perfect square trinomial. Notice the pattern that the first and last terms are perfect squares, and the middle term is twice the product of their roots. The negative middle terms tell you the operation in the binomial. So $16m^2 - 8m + 1 = (4m - 1)^2 = 0$. Setting the binomial equal to 0, $4m - 1 = 0$ giving you $m = \frac{1}{4}$, a double root.

(3) Solve the equation $48x^2 - 10x - 3 = 0$. The answer is $x = \frac{3}{8}, -\frac{1}{6}$.

Here's a perfect candidate for the box method. Put the first and last terms in the box. Multiplying the first and last terms, you get $-144x^2$. You want two factors of this product that add up to the middle term. It'll help to list the possible candidates (without signs, for now): $x \cdot 144x, 2x \cdot 72x, 3x \cdot 48x, 4x \cdot 36x, 6x \cdot 24x, 8x \cdot 18x, 9x \cdot 16x, 12x \cdot 12x$. The pair of factors with a difference of $10x$ is $8x \cdot 18x$. You want $-18x$ and $+8x$. Put those terms in the empty spaces, and then write the GCFs to the side and below.

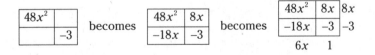

The factorization is picked off the side and bottom: $48x^2 - 10x - 3 = (8x - 3)(6x + 1)$. Setting the factors equal to 0, $x = \frac{3}{8}, -\frac{1}{6}$.

(4) Solve the equation $\frac{1}{6}x^2 + \frac{2}{3}x = 2$. The answer is: $x = 2, x = -6$.

First set the equation equal to 0: $\frac{1}{6}x^2 + \frac{2}{3}x - 2 = 0$.

Now, to get rid of the fractions, multiply each term by 6. This gives you the trinomial $x^2 + 4x - 12 = 0$. This factors to $(x + 6)(x - 2) = 0$. The zero product property gets you to the two solutions: $x = 2$ and $x = -6$.

5 Solve $x^2 - 10 = 2x$. The answer is $x = 1 \pm \sqrt{11}$.

Get 0 on one side first: $x^2 - 2x - 10 = 0$. This equation doesn't factor, so use the quadratic formula to solve:

$$x = \frac{-(-2) \pm \sqrt{(-2)^2 - 4(1)(-10)}}{2(1)} = \frac{2 \pm \sqrt{4 + 40}}{2} = \frac{2 \pm \sqrt{44}}{2} = \frac{2 \pm 2\sqrt{11}}{2} = 1 \pm \sqrt{11}$$

6 Solve $7x^2 - x + 2 = 0$. The answer is "no real solution."

This equation also doesn't factor, so use the quadratic formula to solve:

$$x = \frac{-(-1) \pm \sqrt{(-1)^2 - 4(7)(2)}}{2(7)} = \frac{1 \pm \sqrt{1 - 56}}{14} = \frac{1 \pm \sqrt{-55}}{14}$$

The negative sign under the square root tells you that the solution is not a real number — no real solution exists. The solution involves complex numbers, which are covered in Chapter 11.

7 Solve $x^2 - 4x - 7 = 0$ by completing the square. The answer is $x = 2 \pm \sqrt{11}$.

This time you're asked to complete the square. Begin by adding the 7 to both sides: $x^2 - 4x = 7$. Take half of –4 and square it, and add that to both sides of the equation: $x^2 - 4x + 4 = 11$. Now factor the trinomial: $(x - 2)^2 = 11$. Take the square root of both sides: $\sqrt{(x - 2)^2} = \pm\sqrt{11}$ becomes $x - 2 = \pm\sqrt{11}$. Add the 2 to both sides: $x = 2 \pm \sqrt{11}$.

8 Solve $-2.31x^2 - 4.2x + 6.7 = 0$. The solutions are approximately -2.84 and 1.02.

There's really no way to make the numbers nicer — other than multiplying through by –100. Instead, just reach for your calculator and plug away at the quadratic formula.

$$x = \frac{-(-4.2) \pm \sqrt{(-4.2)^2 - 4(-2.31)(6.7)}}{2(-2.31)} = \frac{4.2 \pm \sqrt{17.64 + 61.908}}{-4.62} = \frac{4.2 \pm \sqrt{79.548}}{-4.62}$$

The two answers show as about -2.839603582 and 1.021421764. Round to the desired number of places.

9 Solve the equation $2x^3 + 3x^2 - 18x + 8 = 0$. The zeros are $x = -4, \frac{1}{2}, 2$.

This third-degree equation has, at most, three real roots. The two changes in sign in $f(x)$ show two or zero positive roots, and the one change in sign in $f(-x)$ shows one negative root.

The list of possible rational zeros is: $\pm 1, \pm 2, \pm 4, \pm 8, \pm \frac{1}{2}$.

Start off by testing $x = 2$.

```
2| 2   3  -18   8
  |     4   14  -8
    2   7   -4   0
```

The depressed polynomial is $2x^2 + 7x - 4$, which factors to $(2x-1)(x+4)$, which tells you that the other two roots are $x = \frac{1}{2}$ and $x = -4$.

(10) Solve the equation $4x^4 + x^3 - 108x - 27 = 0$. The roots are $x = -\frac{1}{4}, 3$.

This polynomial is a candidate for factoring by grouping. The first two terms are divisible by x^3 and the last two are divisible by -27. Doing the first factoring, and then factoring out the common factor of the two resulting terms, $x^3(4x+1) - 27(4+1) = (4x+1)(x^3-27) = 0$. The second factor is the difference of perfect cubes, so you get $(4x+1)(x-3)(x^2+3x+9) = 0$. The first two factors give you the roots $x = -\frac{1}{4}, 3$. The last factor is a trinomial that doesn't factor; using the quadratic formula results in nonreal roots.

(11) Solve the equation $x^3 + 7x^2 + 13x + 4 = 0$. The answers are $x = -4$ and $x = \frac{-3 \pm \sqrt{5}}{2}$.

This cubic polynomial has a maximum of three real roots. None of them are positive and three or one of them is negative. The list of possibilities this time (ignoring all the positives) is: -1, -2, and -4.

Start off by testing -4.

```
-4| 1    7    13    4
  |     -4   -12   -4
   ----------------------
     1    3     1    0
```

The depressed polynomial $x^2 + 3x + 1$ doesn't factor, but the quadratic formula reveals that the last two solutions are $x = \frac{-3 \pm \sqrt{5}}{2}$.

(12) Find the roots of the equation $x^4 + 10x^3 + 38x^2 + 66x + 45 = 0$. The roots are $x = -3$ (double root) and $x = -2 \pm i$.

This fourth-degree polynomial has no positive roots and 4, 2, or 0 negative roots. The list of possibilities to pick from is: $-1, -3, -5, -9, 15, -45$.

Start off with $x = -3$.

```
-3| 1   10    38    66    45
  |      -3   -21   -51   -45
   -------------------------------
     1    7    17    15     0
```

This time, when you test it again, it works.

```
-3| 1    7    17    15
  |      -3   -12   -15
   ------------------------
     1    4     5    0
```

You're left with the depressed polynomial $x^2 + 4x + 5$ which doesn't factor, but you can use the quadratic formula to find that the last two roots are complex: $x = -2 \pm i$.

13 Find the lowest order polynomial with leading coefficient as 1 that has –3, –2, 4, and 6 as its roots. The answer is $x^4 - 5x^3 - 20x^2 + 60x + 144$.

Use the factor theorem to help you figure this one out. If $x = -3$, then $x + 3$ is one of the factors. Similarly, if $x = -2$, then $x + 2$ is a factor; if $x = 4$, then $x - 4$ is a factor; and if $x = 4$, then $x - 6$ is a factor. If you take all the factors and multiply them, you get $(x+3)(x+2)(x-4)(x-6))$. FOIL the first two binomials to get $x^2 + 5x + 6$ and the second two binomials to get $x^2 - 10x + 24$. Multiply your way through those two polynomials: $x^4 - 5x^3 - 20x^2 + 60x + 144$.

14 Find the lowest order polynomial with leading coefficient as 1 that has 2, $4 + 3i$, and $4 - 3i$ as its roots. The answer is $x^3 - 10x^2 + 41x - 50$.

This time the factors are $x - 2$, $x - 4 - 3i$, and $x - 4 + 3i$. In cases like these, it's easier to multiply the complex numbers first. When you do that, you end up with the trinomial $x^2 - 8x + 25$. Now multiply that by the binomial to end up with the polynomial: $x^3 - 10x^2 + 41x - 50$.

15 Factor the polynomial $6x^4 - 7x^3 - 18x^2 + 13x + 6$. The answer is $(x-2)(x-1)(3x+1)(2x+3)$.

You're still using the factor theorem, but this time you have to find the roots first. The roots are $x = 2, 1, -\frac{1}{3}, -\frac{3}{2}$. This means that $x - 2$, $x - 1$, $x + \frac{1}{3}$, and $x + \frac{3}{2}$ are your factors. You can get rid of those fractions by multiplying both terms of the factor by the LCD. In other words, multiply $x + \frac{1}{3}$ by 3 and $x + \frac{3}{2}$ by 2. This finally gives you $(x-2)(x-1)(3x+1)(2x+3)$.

16 Factor the polynomial $x^4 + 10x^3 + 38x^2 + 66x + 45$. The answer is $\left(x^2 + 4x + 5\right)(x+3)^2$.

This is the same expression that appears in Question 12. It has two non-real roots: $x = -2 \pm i$ and $x = -3$, a double root. This means your factors are $(x+2+i)(x+2-i)(x+3)(x+3)$. You multiply out the two non-real roots to come up with a polynomial factor and get $\left(x^2 + 4x + 5\right)(x+3)^2$.

17 Graph $f(x) = x^4 + 2x^3 - 13x^2 - 14x + 24$. See the following graph for the answer.

The polynomial factors: $f(x) = (x+4)(x+2)(x-1)(x-3)$, which gives you the x-intercepts at $x = -4, -2, 1, 3$. Mark those on the graph first. Then find the y-intercept: $y = 24$. The leading coefficient test tells you that both ends of this graph point up. Here's the graph:

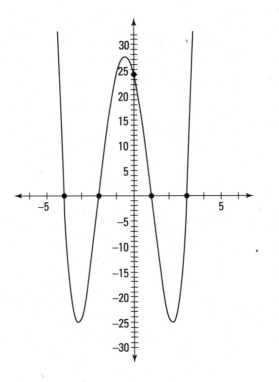

(18) Graph $f(x) = 6x^4 - 7x^3 - 18x^2 + 13x + 6$. See the following graph for the answer.

You found the roots for this polynomial in Question 15: $x = 2, 1, -\frac{1}{3}, -\frac{3}{2}$. Mark those on the graph first. Then find that $(0, 6)$ is the y-intercept. The leading coefficient test tells you that both ends of this graph point up. Here's the graph:

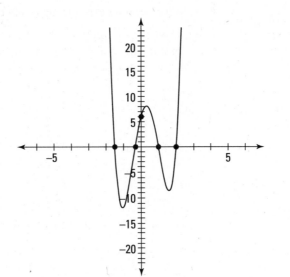

(19) Graph $f(x) = 12x^4 + 13x^3 - 20x^2 + 4x$. See the following graph for the answer.

Factoring, you have $f(x) = x(x+2)(4x-1)(3x-2)$ which tells you that the x-intercepts are when $x = 0, -2, \frac{1}{4}, \frac{2}{3}$. This polynomial has no constant, so $y = 0$ is the y-intercept. This graph crosses at the origin. The leading coefficient test tells you that both ends of the graph point up. Here's the graph:

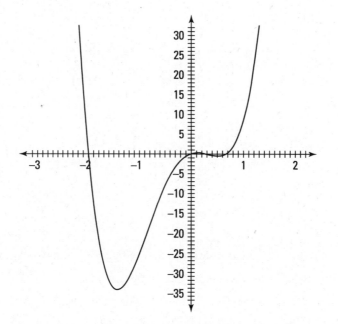

(20) Graph $f(x) = x^4 + 10x^3 + 38x^2 + 66x + 45$. See the following graph for the answer.

This is the same as Question 12 where there are roots of $x = -3$ (as a double root); the other roots are imaginary. The graph will bounce at the double root. The y-intercept is 45. The leading coefficient test tells you that both ends of the graph point up. Here's the graph:

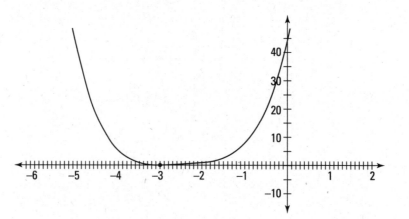

Chapter **5**

Exponential and Logarithmic Functions

E xponential growth is simply the idea that something gets bigger and bigger (or, in the case of exponential decay, smaller and smaller) very fast. Exponential and logarithmic functions can be used to describe growth or decay. These functions have many practical applications, such as determining population growth, measuring the acidity of a substance, and calculating compound interests and loan payments. In addition, they're central to many concepts in calculus and other mathematical arenas.

In this chapter, you practice solving equations, simplifying expressions, and graphing exponential and logarithmic functions. In addition, you practice manipulating functions to solve equations and practically applying the concepts to word problems.

Working with Exponential Functions

Exponential functions are functions in which the variable is in the exponent. When the base of the exponent is greater than 1, the function can grow very quickly, and when it's less than 1, it gets small really fast.

The exponent serves as the power of the expression. The base can be any positive constant except 1, including a special constant that mathematicians and scientists define as e. This irrational constant, e, has a value of approximately 2.7183, and it's extremely useful in exponential and logarithmic expressions.

TECHNICAL STUFF

The basic rules of exponents are used when solving exponential equations or working with exponential expressions:

$$c^a \cdot c^b = c^{a+b} \qquad\qquad \left(c^a\right)^b = c^{a \cdot b}$$

$$\left(\frac{c}{d}\right)^a = \frac{c^a}{d^a}, \ d \neq 0 \qquad\qquad (c \cdot d)^a = c^a \cdot d^a$$

$$\frac{c^a}{c^b} = c^{a-b}, \ c \neq 0 \qquad\qquad c^0 = 1, \ c \neq 0$$

$$c^{-a} = \frac{1}{c^a}, \ c \neq 0 \qquad\qquad c^a = c^b \leftrightarrow a = b, \ c \neq 0,1$$

When graphing exponential equations, it's important to recall the rules for transforming graphs (see Chapter 3 for a refresher).

EXAMPLE

Q. Solve for x in $8^{4x+12} = 16^{2x+5}$.

A. $x = -4$. First, in order to utilize the rules of exponents, it's helpful if both expressions have the same base. So, knowing that $8 = 2^3$ and $16 = 2^4$, you can rewrite both sides of the equation with a base of 2: $\left(2^3\right)^{4x+12} = \left(2^4\right)^{2x+3}$. Using the rule involving raising a power to a power, you have: $2^{3(4x+12)} = 2^{4(2x+3)}$. Now that your bases are the same, you can set your exponents equal to each other (again using properties of exponents): $3(4x+12) = 4(2x+5)$. Next, you can distribute and solve algebraically for x: $12x + 36 = 8x + 20$ becomes $4x = -16$ and $x = -4$.

Q. Sketch the graphs of (A) $y = 2^x$, (B) $y = 2^x + 1$, (C) $y = 2^{x+3}$, (D) $y = 2^{-x}$, and (E) $y = -2^x$, all on the same set of axes.

A. Graphs B–E are all transformations of the first graph, Graph A (see Chapter 3 for a review of transformations of graphs). By adding 1 to Graph A, the result is Graph B, a shift up of 1 unit. By adding 3 to the exponent of Graph A, the result is Graph C, shifted 3 units to the left. Graph D is the result of making the exponent negative, which results in a reflection over the y-axis, and Graph E, created by negating the entire function, results in the reflection of the graph over the x-axis. See the following resulting graphs.

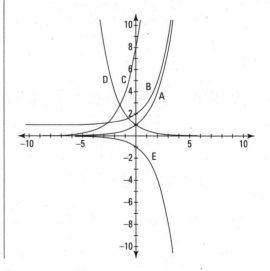

1 Solve for x in $27^{x+3} = 81^{3x-9}$.

2 Solve for x in $e^{2x-4} = e^{6x+8}$.

 3 Solve for x in $\left(5^{2x} - 1\right)\left(25^x - 125\right) = 0$.

 4 Solve for x in $3 \cdot 9^x - 8 = -7$.

5 Sketch the graph of $y = -3^x - 4$.

 6 Sketch the graph of $y = -3e^{x-2}$.

Eagerly Engaging Edgy Logarithmic Solutions

Just as multiplying by the reciprocal is another way to write division, *logarithms* are simply another way to write exponential expressions. Exponential and logarithmic functions are inverses of each other. In other words, logarithmic functions can be used when dealing with exponential functions and vice versa. Logarithms are extremely helpful for an immense number of practical applications. In fact, before the invention of computers, logarithms were the only way to perform many complex computations in physics, chemistry, astronomy, and engineering.

For solving and graphing logarithmic functions (logs), remember this inverse relationship and you'll be sawing . . . er, solving logs in no time! Here's this relationship in equation form:

$$b^y = x \leftrightarrow \log_b x = y$$

TECHNICAL STUFF This equivalence demands that b is greater than 0 and not equal to 1. Observe that $x = b^y > 0$.

Just as with exponential functions, the base can be any positive number except 1, including e. In fact, a base of e is found so frequently in science and calculus that $\log_e$ has its own special symbol or designation: ln. Thus, $\log_e x = \ln x$. These logs are called *natural logs*.

Similarly, $\log_{10}$ is so commonly used that it's often just written as log (without the written base). And these logs have the special name *common logs*.

WARNING Recall the review of domain in Chapter 3. The domain for the basic logarithm $y = \log_b x$ is $x > 0$. Therefore, when you're solving logarithmic functions, it's important to check for extraneous roots (review Chapter 1).

Here are more properties that are true for all logarithms:

$$\log_b 1 = 0 \qquad\qquad \log_b b = 1$$

$$\log_b (a \cdot c) = \log_b a + \log_b c \qquad \log_b \left(\frac{a}{c}\right) = \log_b a - \log_b c$$

$$\log_b a^c = c \cdot \log_b a \qquad\qquad \log_b b^x = x$$

$$b^{\log_b x} = x,\ x > 0 \qquad\qquad \log_b a = \frac{\log_c a}{\log_c b}$$

$$\log_b a = \log_b c \to a = c$$

Using these properties, simplifying logarithmic expressions and solving logarithmic equations become easier tasks.

EXAMPLE

Q. Rewrite the following logarithmic expression to a single log: $3\log_5 x + \log_5 (2x-1) - 2\log_5 (3x+2)$.

A. $\log_5 \dfrac{x^3(2x-1)}{(3x+2)^2}$

Using the properties of logs, begin by rewriting the coefficients as exponents: $3\log_5 x = \log_5 x^3$ and $2\log_5 (3x+2) = \log_5 (3x+2)^2$. Next, rewrite the addition of the first two logs as the log of the product of two functions: $\log_5 x^3 + \log_5 (2x-1) = \log_5 x^3(2x-1)$. Last, rewrite the difference of these two logs as the log of the quotient: $\log_5 x^3(2x-1) - \log_5 (3x+2)^2 = \log_5 \dfrac{x^3(2x-1)}{(3x+2)^2}$.

Q. Sketch the graphs of (A) $y = \log x$, (B) $y = 1 + \log(x-2)$, (C) $y = \log(x+3)$, and (D) $y = -\log x$, all on the same set of axes.

A. First, in the following figure, you can see that Graphs B–D are transformations of Graph A. Graph B is a shift of 1 up, Graph C is a shift of 3 to the left, and Graph D is a reflection of Graph A over the x-axis. Secondly, you'll find that these are all inverses of Graphs A–D in the preceding section on solving exponential functions. Another way to graph logarithms is to change the log to an exponential function. Using the properties of logarithms, find the inverse function by switching x and y, graph the inverse, and reflect every point over the line $y = x$. For a review of inverses, see Chapter 3.

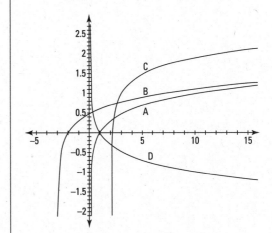

7 Rewrite the given expression as a single logarithm:

$$\ln 4x + 3\ln(x-2) - 2(\ln 2x + \ln(3x-4)).$$

8 Solve for x in $\log_7 4 + \log_7(x+4) - 2\log_7 2 =$

$$\log_7(x-2) + \frac{1}{2}\log_7 9.$$

9 Solve for x in $\ln x + \ln(2x-1) = \ln 1$.

10 Find $\log_b(48b)$ if $\log_b 2 = 0.36$ and $\log_b 3 = 0.56$.

11   Sketch the graph of $y = -3 + \log(x+2)$.

 12   Sketch the graph of $y = \ln(x-2) + 4$.

Making Exponents and Logs Work Together

In this section, you see how the log and exponential functions can be used to solve various problems. By keeping in mind the inverse relationship, $b^y = x \leftrightarrow \log_b x = y$, you can solve some interesting and even challenging problems.

TIP

A helpful key to remember when solving equations using exponents and logs is that if the variable is in the exponent, you can convert the equation into logarithmic form. This is especially helpful if you use *natural log* (ln) or the *common log* ($\log_{10} x$), often referred to as just $\log x$, because you can plug the expression into your calculator to get a decimal approximation of the solution.

WARNING

One pitfall to avoid when manipulating logs relates to the products and quotients of logs. Remember: $\log 18 - \log 3 = \log\left(\dfrac{18}{3}\right) = \log 6$, not $\dfrac{\log 18}{\log 3}$. These are entirely different expressions. In fact, if you plug them into your calculator, you can see that $\log\left(\dfrac{18}{3}\right) = \log 6 \approx 0.778$, whereas $\dfrac{\log 18}{\log 3} \approx 2.631$. The same can be said for products and logs: $\log 6 + \log 7 = \log(6 \cdot 7) = \log 42$, not $\log(6+7)$ nor $(\log 6)(\log 7)$.

Q. Solve for x in $\log(50x + 250) - \log x = 2$.

EXAMPLE **A.** $x = 5$. Start by combining the logs as a quotient: $\log\dfrac{(50x + 250)}{x} = 2$. Next, rewrite in exponential form (remember that $\log x$ means $\log_{10} x$): $10^2 = \dfrac{50x + 250}{x}$. After cross-multiplying, you can then solve algebraically: $100x = 50x + 250$ giving you $50x = 250$, and then $x = 5$.

Q. Solve for x in $3^x = 2^{x+2}$.

A. $x = \dfrac{\ln 4}{\ln\left(\dfrac{3}{2}\right)}$. First, recognize that the variable is in the exponent of each term, so you can easily remedy that by taking either log or ln of both sides. In this case, ln is used, but it really doesn't make a difference. So $3^x = 2^{x+2}$ becomes $\ln 3^x = \ln 2^{x+2}$. Then, you can use properties of logarithms to solve. Start by changing the exponents to coefficients: $x\ln 3 = (x + 2)\ln 2$. Using algebra, you can distribute the $\ln 2$ across $(x + 2)$ to make the equation: $x\ln 3 = x\ln 2 + 2\ln 2$. Get the terms with the variable on the same side by subtracting $x\ln 2$ from each side: $x\ln 3 - x\ln 2 = 2\ln 2$. Then factor to get: $x(\ln 3 - \ln 2) = 2\ln 2$. Divide each side by the multiplier of x: $x = \dfrac{2\ln 2}{\ln 3 - \ln 2}$. Last, use log rules to simplify the numerator and denominator and get

$$x = \dfrac{\ln 2^2}{\ln\left(\dfrac{3}{2}\right)} = \dfrac{\ln 4}{\ln\left(\dfrac{3}{2}\right)}.$$

13 Solve for x in $\log(x + 6) - \log(x - 3) = 1$.

14 Solve for x in $3^x = 5$.

15   Solve for x in $4^x - 4 \cdot 2^x = -3$.

16   Solve for x in $3^x = 5^{2x-3}$.

Using Exponents and Logs in Practical Applications

When will I ever use this? Well, in addition to being used in mathematics courses, exponential functions actually have many practical applications. Common uses of exponential functions include figuring compound interest, computing population growth, and doing radiocarbon dating (no, that's not some new online matchmaking system). Log functions are found in sound and chemical formulas.

TECHNICAL STUFF

Here are formulas for interest rate and half-life:

>> **Compound interest formula:** $A = P\left(1 + \dfrac{r}{n}\right)^{nt}$ where A is the total amount after t time in years compounded n times per year if P dollars are invested at annual interest rate r.

>> **Continuous compound interest formula:** $A = Pe^{rt}$, where A is the total amount after t time in years if P dollars are invested at interest rate r with interest compounded continuously throughout the year.

>> **Formula for the remaining mass of a radioactive element:** $M(x) = c \cdot 2^{-x/h}$, where $M(x)$ is the mass at the time x, c is the original mass of the element, and h is the half-life of the element.

EXAMPLE

Q. If you deposit $600 at 5.5% interest compounded continuously, what will your balance be in 10 years?

A. $1,039.95. Because this is continuous compound interest, you use the formula $A = Pe^{rt}$ when you're solving for A: $A = 600e^{(0.055)(10)}$. Plugging this into a calculator, you get approximately $1,039.95.

Q. How old is a piece of bone that has lost 60 percent of its carbon-14? (The half-life of carbon-14 is 5,730 years.)

A. Approximately 7,575 years old. You can figure out this problem using the formula for the remaining mass of an element. First, because 60 percent of the carbon-14 is gone, the mass of carbon remaining is 40 percent, so you can write the present mass as $0.40c$. Therefore, the equation will be: $0.40c = c \cdot 2^{-x/5730}$. You can start solving this by dividing each side by c: $0.40 = 2^{-x/5730}$. Taking the natural log of both sides allows you to move the variable from the exponent position: $\ln 0.40 = \ln\left(2^{-x/5730}\right)$ becomes $\ln 0.40 = -\dfrac{x}{5730} \cdot \ln 2$. Solving for x, you have $-\dfrac{5730}{\ln 2} \cdot \ln 0.40 = x$. And your calculator tells you that $x \approx 7,575$ years.

---

17 If you deposit $3,000 at 8-percent interest per year compounded quarterly, in approximately how many years will the investment be worth $10,500?

18 The half-life of Krypton-85 is 10.4 years. How long will it take for 600 grams to decay to 15 grams?

19 The deer population in a certain area in year t is approximately $P(t) = \dfrac{3{,}000}{1 + 299e^{-0.56t}}$.

When will the deer population reach 2,000?

20 If you deposit $20,000 at 6.5-percent interest compounded continuously, how long will it take for you to have $1 million?

Answers to Problems on Exponential and Logarithmic Functions

Following are the answers to problems dealing with exponential and logarithmic functions.

(1) Solve for x in $27^{x+3} = 81^{3x-9}$. The answer is $x = 5$.

First, rewrite 27 as 3^3 and 81 as 3^4. Rewrite the equation and simplify to get $\left(3^3\right)^{x+3} = \left(3^4\right)^{3x-9}$ and then $3^{3x+9} = 3^{12x-36}$. Now that the bases are the same, set the two exponents equal to each other: $3x + 9 = 12x - 36$, and then solve for x: $-9x = -45$ and $x = 5$.

(2) Solve for x in $e^{2x-4} = e^{6x+8}$. The answer is $x = -3$.

Start by setting the exponents equal to each other: $2x - 4 = 6x + 8$; then solve for x: $-4x = 12$ and $x = -3$.

(3) Solve for x: $\left(5^{2x} - 1\right)\left(25^x - 125\right) = 0$. The answer is $x = 0, \dfrac{3}{2}$.

Using the fact that $25 = 5^2$, replace 25^x with 5^{2x} to get $\left(5^{2x} - 1\right)\left(5^{2x} - 125\right) = 0$. Next, set each factor equal to 0 using the zero product property (see Chapter 4 for a review) and solve: First, $5^{2x} - 1 = 0$, which implies $5^{2x} = 1$ and, because anything to the power of 0 equals 1, $5^{2x} = 5^0$. Therefore, $2x = 0$ and thus $x = 0$. Second, $5^{2x} - 125 = 0$, which implies $5^{2x} = 125$, and because 125 is equal to 5^3, rewrite the second equation as $5^{2x} = 5^3$. Set the exponents equal to each other, $2x = 3$, and solve to get $x = \dfrac{3}{2}$. Both solutions work.

(4) Solve for x in $3 \cdot 9^x - 8 = -7$. The answer is $x = -\dfrac{1}{2}$.

Start by isolating the exponential expression: $3 \cdot 9^x = 1$. Dividing, $9^x = \dfrac{1}{3}$. Next, replace 9^x with 3^{2x} and $\dfrac{1}{3}$ with 3^{-1}, so $3^{2x} = 3^{-1}$. Set the exponents equal to each other: $2x = -1$, and solve for x to get $x = -\dfrac{1}{2}$.

(5) Sketch the graph of $y = -3^x - 4$. See the graph for the answer.

The y-intercept is $(0, -5)$. The graph of this function is the basic exponential graph of $y = 3^x$ reflected across the x-axis first and then shifted 4 units down.

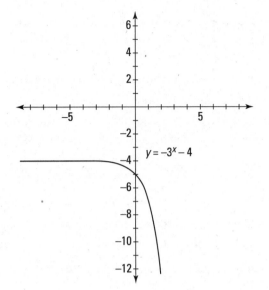

$y = -3^x - 4$

(6) Sketch the graph of $y = -3e^{x-2}$. See the graph for the answer.

The y-intercept is approximately $(0, -0.406)$. The graph of this function is the basic exponential graph of $y = e^x$ shifted 2 units to the right, reflected across the x-axis, and followed with a steepening by a factor of 3.

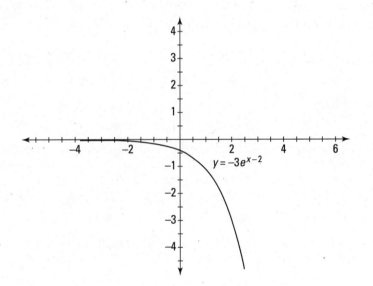

$y = -3e^{x-2}$

(7) Rewrite the given expression as a single logarithm: $\ln 4x + 3\ln(x-2) - 2(\ln 2x + \ln(3x-4))$. The answer is $\ln\left[\dfrac{(x-2)^3}{x(3x-4)^2}\right]$.

Begin by distributing the -2 and rewriting coefficients as exponents: $\ln 4x + 3\ln(x-2) - 2\ln 2x - 2\ln(3x-4)$ becomes $\ln 4x + \ln(x-2)^3 - \ln(2x)^2 - \ln(3x-4)^2$. Next, rewrite the first two logarithms as a single product and the last two terms as a product to get $\ln 4x(x-2)^3 - \ln 4x^2(3x-4)^2$. Finally, write the difference of logarithms as the log of a quotient and reduce the fraction: $\ln\left[\dfrac{4x(x-2)^3}{4x^2(3x-4)^2}\right] = \ln\left[\dfrac{(x-2)^3}{x(3x-4)^2}\right]$.

(8) Solve for x in: $\log_7 4 + \log_7(x+4) - 2\log_7 2 = \log_7(x-2) + \dfrac{1}{2}\log_7 9$. The answer is $x = 5$.

The first step is to write the coefficients as exponents: $\log_7 4 + \log_7(x+4) - \log_7 2^2 = \log_7(x-2) + \log_7 9^{1/2}$. Next, simplify the powers created and rewrite the sums and differences of logs as the logs of products and quotients: $\log_7\left[\dfrac{4(x+4)}{4}\right] = \log_7(x-2)3$. Using the rules of logarithms, set $\dfrac{4(x+4)}{4} = (x-2)3$. Simplify and solve: $x + 4 = 3x - 6$ becomes $-2x = -10$, giving you $x = 5$.

(9) Solve for x in $\ln x + \ln(2x-1) = \ln 1$. The answer is $x = 1$.

The domain requirements from the two log terms in the equation require $x > \dfrac{1}{2}$. Rewriting the sum of natural logs as the log of a product, you get: $\ln x(2x-1) = \ln 1$. Then, using rules of logarithms, set $x(2x-1) = 1$ and solve for x: $2x^2 - x = 1$ becomes $2x^2 - x - 1 = 0$. Factoring the quadratic, $(2x+1)(x-1) = 0$ and, using the zero product property, set each factor equal to 0. The solutions are $x = -\dfrac{1}{2}$ and $x = 1$, but $-\dfrac{1}{2}$ is not in the domain, so it's extraneous, and the only solution is $x = 1$.

(10) Find $\log_b(48b)$ if $\log_b 2 = 0.36$ and $\log_b 3 = 0.56$. The answer is 3.

Start by expanding the logarithm into the sum of two logs: $\log_b 48 + \log_b b$. Next, factoring the 48 into $16 \cdot 3 = 2^4 \cdot 3$, rewrite the first log: $\log_b 2^4 + \log_b 3 + \log_b b$. Then, write the exponent as a coefficient: $4\log_b 2 + \log_b 3 + \log_b b$. Using the rules for logs, $\log_b b = 1$. So now you replace the log terms with their values and simplify: $4(0.36) + 0.56 + 1 = 3$.

(11) Sketch the graph of $y = -3 + \log(x+2)$. See the graph for the answer.

The y-intercept is approximately $(0, -2.699)$. The graph of this function is the basic logarithmic graph of $y = \log x$ shifted 2 units to the left and 3 units down.

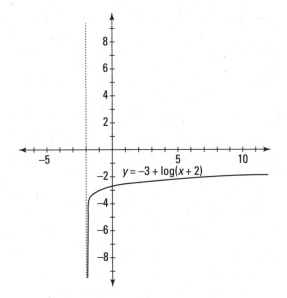

(12) Sketch the graph of $y = \ln(x-2)+4$. See the graph for the answer.

There's a vertical asymptote at $x = 2$. The graph of this function is the basic logarithmic graph of $y = \ln x$ shifted 2 units to the right and 4 units up.

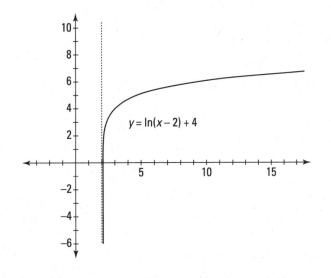

(13) Solve for x in $\log(x+6) - \log(x-3) = 1$. The answer is $x = 4$.

Begin by writing the difference of logs as the log of a quotient: $\log \dfrac{x+6}{x-3} = 1$. Next, rewrite the logarithm in its equivalent exponential form: $10^1 = \dfrac{x+6}{x-3}$. Then, solve for x: $10x - 30 = x + 6$ becomes $9x = 36$ and $x = 4$. You see 4 is indeed a solution.

(14) Solve for x in $3^x = 5$. The answer is $x = \dfrac{\ln 5}{\ln 3}$ or $x \approx 1.465$.

First, take the natural log of both sides: $\ln 3^x = \ln 5$. Then use the power rule to simplify: $x \ln 3 = \ln 5$.

Solving for x: $x = \dfrac{\ln 5}{\ln 3}$ or $x \approx 1.465$.

(15) Solve for x in $4^x - 4 \cdot 2^x = -3$. The answer is $x = \dfrac{\ln 3}{\ln 2} \approx 1.585$ and $x = 0$.

Start by using the fact that $4 = 2^2$ and rewrite the equation as: $2^{2x} - 4 \cdot 2^x = -3$. Add 3 to both sides $2^{2x} - 4 \cdot 2^x + 3 = 0$. Notice that this is the same as the quadratic-like equation $\left(2^x\right)^2 - 4 \cdot 2^x + 3 = 0$. So you can substitute y for 2^x to arrive at $y^2 - 4y + 3 = 0$. Now, you can factor using the new format and solve using the zero product property: $(y-3)(y-1) = 0$; $y = 3$ and $y = 1$. Then, resubstitute 2^x for y: $2^x = 3$ and $2^x = 1$. Taking the natural log of each side, you can solve for x by using the rules of logarithms: $\ln 2^x = \ln 3$ gives you $x \ln 2 = \ln 3$ or $x = \dfrac{\ln 5}{\ln 3}$, your first solution. And then, $\ln 2^x = \ln 1$ becomes $\ln 2^x = 0$, using a rule for logarithms and $x \ln 2 = 0$. So $x = \dfrac{0}{\ln 2} = 0$ your second solution. Both solutions work.

(16) Solve for x in $3^x = 5^{2x-3}$. The answer is $x = -\dfrac{3\ln 5}{\ln 3 - 2\ln 5} \approx 2.277$.

First, take the natural log of both sides: $\ln 3^x = \ln 5^{2x-3}$. Then use properties of logarithms to solve. Start by changing the exponents to coefficients: $x \ln 3 = (2x - 3)\ln 5$. Distribute the $\ln 5$: $x \ln 3 = 2x \ln 5 - 3 \ln 5$. Get the terms with the variable on the same side: $x \ln 3 - 2x \ln 5 = -3 \ln 5$. Then factor $x (\ln 3 - 2 \ln 5) = -3 \ln 5$ and solve for x: $x = -\dfrac{3\ln 5}{\ln 3 - 2\ln 5}$.

(17) If you deposit \$3,000 at 8-percent interest per year, compounded quarterly, in approximately how many years will the investment be worth \$10,500? The answer is approximately 15.82 years.

Using the equation: $A = P\left(1 + \dfrac{r}{n}\right)^{nt}$, where $A = \$10{,}500$, $P = \$3{,}000$, $r = 0.08$, and $n = 4$, you can solve for t: $10{,}500 = 3{,}000\left(1 + \dfrac{0.08}{4}\right)^{4t}$. Simplifying, you have $10{,}500 = 3{,}000(1.02)^{4t}$, and then $3.5 = (1.02)^{4t}$. Using logarithms, $\log 3.5 = \log(1.02)^{4t}$ which is written $\log 3.5 = 4t \log(1.02)$. Solving for t: $\dfrac{\log 3.5}{4 \log 1.02} = t$

which equals approximately 15.82 years.

(18) The half-life of Krypton-85 is 10.4 years. How long will it take for 600 grams to decay to 15 grams? The answer is 55.3 years.

Using the half-life formula: $M(x) = c \cdot 2^{-x/h}$, where $M(x) = 15$ grams, the original mass c is 600 grams, and the half-life h is 10.4. You get $15 = 600 \cdot 2^{-x/10.4}$. Dividing, you get $0.025 = 2^{-x/10.4}$ which can be solved using logarithms: $\log 0.025 = \log 2^{-x/10.4}$ becomes $\log 0.025 = -\dfrac{x}{10.4} \log 2$. Solving for x, you have $-10.4 \left(\dfrac{\log 0.025}{\log 2}\right) = x$, which approximates to 55.3 years.

(19) The deer population in a certain area in year t is approximately $P(t) = \dfrac{3,000}{1+299e^{-0.56t}}$. When will the deer population reach 2,000? The answer is approximately 11.4 years. Here, you simply plug in 2,000 for $P(t)$ and solve for t: $2,000 = \dfrac{3,000}{1+299e^{-0.56t}}$, $2,000\left(1+299e^{-0.56t}\right) = 3,000$, $1+299e^{-0.56t} = 1.5$, $299e^{-0.56t} = 0.5$, $e^{-0.56t} = \dfrac{0.5}{299}$. Now, take the ln of each side: $\ln e^{-0.56t} = \ln\dfrac{0.5}{299}$.

Using two laws of logarithms, the left simplifies: $-0.56t = \ln\dfrac{0.5}{299}$, so $t = \dfrac{\ln\dfrac{0.5}{299}}{-0.56}$ which equals approximately 11.4 years.

(20) If you deposit $20,000 at 6.5-percent interest compounded continuously, how long will it take for you to have $1 million? The answer is approximately 60.2 years.

Using the equation for continuous compound interest, $A = Pe^{rt}$, where the amount A is $1,000,000, the initial investment P is $20,000, and the interest rate r in decimal form is 0.065, you get: $1,000,000 = 20,000e^{0.065t}$. Simplify and use logarithms to solve: $50 = e^{0.065t}$; $\ln 50 = 0.065t$; $\dfrac{\ln 50}{0.065} = t$, which equals approximately 60.2 years.

2

Trig Is the Key: Basic Review, the Unit Circle, and Graphs

IN THIS PART . . .

You should be familiar with the basics of trigonometry from earlier math classes — right triangles, trig ratios, and angles, for example. But your Algebra II course may or may not have expanded on those ideas to prepare you for the direction that pre-calculus is going to take you. For this reason, it's assumed that you've never seen this stuff before. You won't be left behind when continuing your mathematics journey.

This part begins with trig ratios and word problems and then moves on to the unit circle: how to build it and how to use it. You'll some trig equations and make and measure arcs. Graphing trig functions is a major component of pre-calculus, so you'll see how to graph each of the six functions.

IN THIS CHAPTER

» **Working with the six trigonometric ratios**

» **Making use of right triangles to solve word problems**

» **Using the unit circle to find points, angles, and right triangle ratios**

» **Isolating trig terms to solve trig equations**

» **Calculating arc lengths**

Chapter **6**

Basic Trigonometry and the Unit Circle

A h . . . trigonometry, the math of triangles! Invented by the ancient Greeks, trigonometry is used to solve problems in navigation, astronomy, and surveying. Think of a sailor lost at sea. All he has to do is triangulate his position against two other objects, such as two stars, and calculate his position using — you guessed it — trigonometry!

In this chapter, the basics of right triangle trigonometry are reviewed. Then you see how to apply that knowledge to the unit circle, a very useful tool for graphically representing trigonometric ratios and relationships. From there, you can solve trig equations. Finally, these concepts are combined so that you can apply them to arcs. The ancient Greeks didn't know what they started with trigonometry, but the modern applications are endless!

Finding the Six Trigonometric Ratios

In geometry, angles are measured in degrees, with 360° describing a full circle on a coordinate plane. And another measure for angles is *radians*. Radians, from the word radius, are usually designated without a symbol for units; the radian measure is understood. Because both radians and degrees are used often in pre-calculus, you see both in use here.

TIP
To convert radians to degrees and vice versa, you use the fact that $360° = 2\pi$ radians, or $180° = \pi$. Therefore, to convert degrees to radians, just remember this proportion and substitute in what you know: $\dfrac{\theta°}{180} = \dfrac{\theta^R}{\pi}$. The $\theta°$ represents an angle measured in degrees, and the θ^R represents an angle measured in radians.

When solving right triangles or finding all the sides and angles, it's important to remember the six basic trigonometric functions: $\sin\theta$, $\cos\theta$, $\tan\theta$, $\csc\theta$, $\sec\theta$, and $\cot\theta$. And there's also the reciprocal relationships between the functions: $\sin\theta = \dfrac{1}{\csc\theta}$, $\cos\theta = \dfrac{1}{\sec\theta}$, and $\tan\theta = \dfrac{1}{\cot\theta}$.

TIP
One of the most famous acronyms in math is SOHCAHTOA. It helps you remember the first three trigonometric ratios:

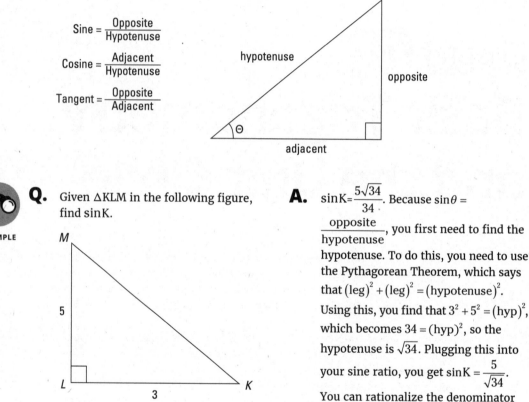

$$\text{Sine} = \frac{\text{Opposite}}{\text{Hypotenuse}}$$

$$\text{Cosine} = \frac{\text{Adjacent}}{\text{Hypotenuse}}$$

$$\text{Tangent} = \frac{\text{Opposite}}{\text{Adjacent}}$$

Q. Given △KLM in the following figure, find $\sin K$.

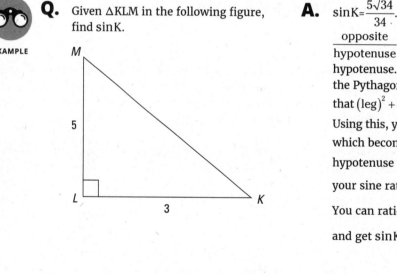

A. $\sin K = \dfrac{5\sqrt{34}}{34}$. Because $\sin\theta = \dfrac{\text{opposite}}{\text{hypotenuse}}$, you first need to find the hypotenuse. To do this, you need to use the Pythagorean Theorem, which says that $(\text{leg})^2 + (\text{leg})^2 = (\text{hypotenuse})^2$. Using this, you find that $3^2 + 5^2 = (\text{hyp})^2$, which becomes $34 = (\text{hyp})^2$, so the hypotenuse is $\sqrt{34}$. Plugging this into your sine ratio, you get $\sin K = \dfrac{5}{\sqrt{34}}$. You can rationalize the denominator and get $\sin K = \dfrac{5\sqrt{34}}{34}$.

Q. Solve △RST, referring to the following figure.

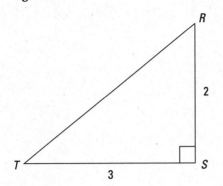

R

2

T

3

S

A. $RT = \sqrt{13}$, $\angle T \approx 33.7°$, $\angle R \approx 56.3°$.

Remember, solving a triangle means finding the measures of all the angles and sides. So you start by using the Pythagorean Theorem to find the hypotenuse: $2^2 + 3^2 = RT^2$, $RT = \sqrt{13}$. Next, use any trigonometric ratio to find an angle. You can use $\sin T = \dfrac{2}{\sqrt{13}}$.

To get the measure of the angle, you use the inverse operation of sin, is $\sin^{-1}\theta$ in your calculator. Thus, you have $\sin^{-1}\left(\dfrac{2}{\sqrt{13}}\right) \approx 33.7°$. Or, if you want to use radians, $\angle T \approx 0.59$. Lastly, using the fact that the angles of a triangle add up to 180°, you can find $\angle R$: $180 - (90 + 33.7) = 180 - 123.7 = 56.3$.

---

1 Find $\cos A$ in △ABC.

2 Solve △DEF.

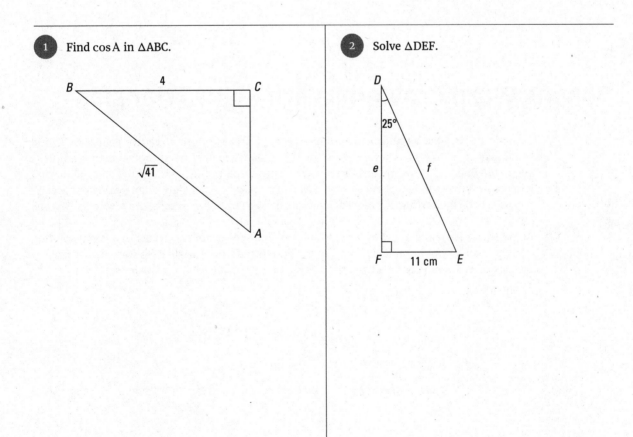

3 Find ∠Q in △QRS(round to the nearest tenth).

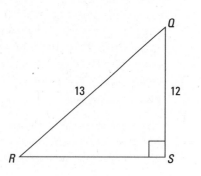

 4 Find the six trigonometric ratios of ∠R in △QRS from Question 3.

Solving Word Problems with Right Triangles

Uh-oh! The dreaded word problems! But trig word problems are a relative snap. You'll find some easy steps to help you through them. First, as with most word problems in math, it is suggested that you draw a picture. That way you can visualize the problem and determine what trig property you can use. Second, remember that these are just right triangles. Therefore, all you have to do is use what you already know about right triangles to solve the problems. Simple!

TIP

Angle of elevation and angle of depression (see Figure 6-1) are two terms that come up often in right triangle word problems. They just refer to whether the angle rises from the horizon — an *angle of elevation* — or falls from the horizon, called an *angle of depression*.

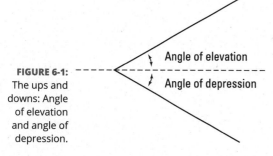

FIGURE 6-1:
The ups and downs: Angle of elevation and angle of depression.

Q. When the sun is at an angle of elevation of 32°, a building casts a shadow that's 152 feet from its base. What is the height of the building?

A. The building is approximately 95 feet tall. Okay, remember your steps. Step one, draw a picture:

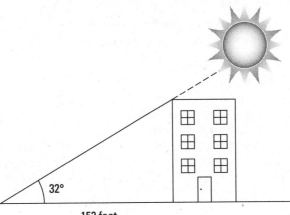

152 feet

Step two, recall what you know about right triangles. Because you want to find the building's height, x, which is opposite the angle, and you have the shadow length, which is adjacent to the angle, you can use the tangent ratio. Setting up your ratio, you get $\tan 32° = \dfrac{x}{152}$, or $x = 152 \cdot \tan 32°$. Using a calculator, you find that the building height is approximately 95 feet.

Q. Two boat captains whose boats are in a straight line from a lighthouse look up to the top of the lighthouse at the same time. The captain of Boat A sees the top of the 40-foot lighthouse from an angle of elevation of 45°, while the captain of Boat B sees the top of the lighthouse from an angle of elevation of 30°. How far are the boats from each other, to the nearest foot?

A. The boats are 29 feet apart.

First, remember to draw a picture.

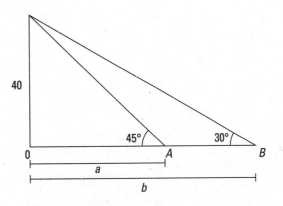

From the picture, you can see that to find the distance between the boats, you need to find the distance that each boat is from the base of the lighthouse and subtract Boat A's distance from the distance of Boat B. Because the angle of elevation is 45° for Boat A, you can set up the trigonometric ratio: $\tan 45° = \dfrac{40}{a}$. Solving for a, you find that the distance from Boat A to the base of the lighthouse is 40 feet (to the nearest foot).

Similarly, you can set up a trigonometric ratio for Boat B's distance: $\tan 30° = \dfrac{40}{a}$. Solving, you get that b = 69 feet. Subtracting these two distances, you find that the distance between the boats is 29 feet. Whew!

---

5 Romero wants to deliver a rose to his girl-friend, Jules, who is sitting on her balcony 24 feet above the street. If Romero has a 28-foot ladder, at what angle must he place the bottom of the ladder to reach his love, Jules?

6 Sam needs to cross a river. He spies a bridge directly ahead of him. Looking across the river, he sees that he's 27° below the bridge from the other side. How far must he walk on his side of the river to reach the bridge if the bridge length is 40 feet?

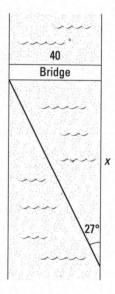

7 Paul, a 6-foot-tall man, is holding the end of a kite string 5 feet above the ground. The string of the kite is 75 feet long at 35° of elevation. Paulette, Paul's 5-foot-tall daughter, is directly underneath the kite. How far above Paulette's head is the kite?

8 To hold up a 100-foot pole, two guide wires are put up on opposite sides of the pole. One wire makes a 36° angle with the ground, and the other makes a 47° angle with the ground. How far apart are the bases of the wires?

Unit Circle and the Coordinate Plane: Finding Points and Angles

The unit circle is a very useful tool in pre-calculus. The information it provides can help you solve problems very quickly. Essentially, the *unit circle* is a circle with a radius (*r*) of one unit, centered on the origin of a coordinate plane. Thinking of the trigonometric ratios you've been dealing with in terms of *x* and *y* values, where *x* is adjacent to the angle, *y* is opposite the angle, and *r* is the hypotenuse, allows you to make a right triangle by using a point on the unit circle and the *x*-axis. This is often called *point-in-plane*, and it results in an alternate definition of the six trigonometric ratios:

$$\sin\theta = \frac{y}{r} \quad \csc\theta = \frac{r}{y}$$

$$\cos\theta = \frac{x}{r} \quad \sec\theta = \frac{r}{x}$$

$$\tan\theta = \frac{y}{x} \quad \cot\theta = \frac{x}{y}$$

When graphing on a coordinate plane, how you measure your angles is important. In pre-calculus, the angle always begins on the positive side of the *x*-axis, called the *initial side*. Any angle in this position is in *standard position*. The angle can extend to anywhere on the plane, ending on what's called the *terminal side*. Any angles that have different measures but have the same terminal side are called *co-terminal angles*. These can be found by adding or subtracting 360° or 2π to any angle.

REMEMBER

From the initial side, an angle that moves in the counterclockwise direction has a *positive measure*, and an angle that moves in the clockwise direction has a *negative measure*.

EXAMPLE

Q. Find three co-terminal angles of 520°.

A. Sample answer: 160°, –200°, and 880°, but other answers are possible. To get these angles, you simply add or subtract multiples of 360° from 520°. $520° - 360° = 160°$; $520° - 2(360°) = -200°$; and $520° + 360° = 880°$.

Q. Evaluate the six trigonometric ratios of an angle whose terminal side goes through the point $(2, -3)$.

A. $\sin\theta = -\dfrac{3\sqrt{13}}{13}, \cos\theta = \dfrac{2\sqrt{13}}{13}, \tan\theta = -\dfrac{3}{2},$
$\csc\theta = -\dfrac{\sqrt{13}}{3}, \sec\theta = \dfrac{\sqrt{13}}{2}, \cot\theta = -\dfrac{2}{3}.$

Start by finding the radius using the Pythagorean Theorem: $2^2 + (-3)^2 = r^2$, $r^2 = 13, r = \sqrt{13}$. Then, simply plug the known values into the trigonometric ratios given: $x = 2$, $y = -3$, and $r = \sqrt{13}$. Don't forget to rationalize any radicals in the denominator!

$$\sin\theta = \frac{-3}{\sqrt{13}} = -\frac{3\sqrt{13}}{13}, \cos\theta = \frac{2}{\sqrt{13}} = \frac{2\sqrt{13}}{13},$$

$$\tan\theta = -\frac{3}{2}, \csc\theta = -\frac{\sqrt{13}}{3}, \sec\theta = \frac{\sqrt{13}}{2},$$

$$\cot\theta = -\frac{2}{3}.$$

9 Find three co-terminal angles of $\dfrac{\pi}{5}$.

10 Find two positive co-terminal angles of –775°.

Evaluate the six trigonometric ratios of an angle whose terminal side goes through the point $(3, 4)$.

 Evaluate the six trigonometric ratios of an angle whose terminal side goes through the point $(-5, -7)$.

 Evaluate the six trigonometric ratios of an angle whose terminal side goes through the point $\left(-2, 2\sqrt{3}\right)$.

 Evaluate the six trigonometric ratios of an angle whose terminal side goes through the point $\left(6, -3\sqrt{5}\right)$.

Finding Ratios from Angles on the Unit Circle

Well, isn't that special? Yes, it *is* special — special right triangles that is! Remember your geometry teacher drilling in 30-60-90 and 45-45-90 triangles? Well, they're back! And with good reason, because they're very common in pre-calculus, and they're the foundation of the unit circle. Using these two special triangles, you can find the specific trig values that you see on the completed unit circle in Figure 6-2.

REMEMBER

One important point to remember about the unit circle is that the radius is 1. Therefore, the hypotenuse of any right triangle drawn from a point to the x-axis is 1. Thus, for any point, (x, y), you know that $x^2 + y^2 = 1$.

Recalling 30-60-90 triangles, the sides are in the ratio of $1 : \sqrt{3} : 2$. Therefore, if you want the hypotenuse to be 1, as it is in the unit circle, divide each side by 2. Similarly, the sides of 45-45-90 triangles are in the ratio of $1 : 1 : \sqrt{2}$. Converting to a unit circle, the values are $\dfrac{\sqrt{2}}{2} : \dfrac{\sqrt{2}}{2} : 1$.

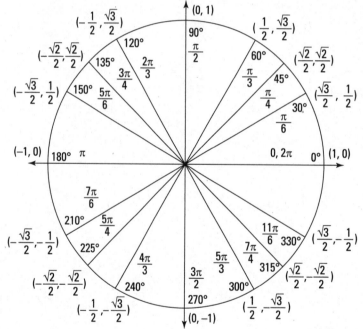

FIGURE 6-2:
The whole unit circle.

Now, using the point-in-plane definition, the six trigonometric ratios are easy to find. In fact, because the hypotenuse is now 1, $\sin\theta = \dfrac{y}{r}$ becomes $\sin\theta = y$. Similarly, $\cos\theta = \dfrac{x}{r}$ becomes $\cos\theta = x$. Thus, any point on the unit circle has the coordinates $(\cos\theta, \sin\theta)$. Imagine the possibilities!

If you don't have a unit circle handy, you can always use *reference angles* to find your solutions. A reference angle has the same terminal side as the angle in question and is the angle between the *x*-axis and that terminal side. Reference angles are different for each quadrant (see Figure 6-3). If the original angle is θ, then the reference angle in Quadrant I is $\theta' = \theta$. In Quadrant II, the reference angle is $\theta' = 180° - \theta$. For Quadrant III, the reference angle is $\theta' = \theta - 180°$. Lastly, Quadrant IV's reference angle is $\theta' = 360° - \theta$.

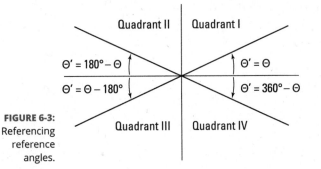

FIGURE 6-3: Referencing reference angles.

Q. Evaluate the six trigonometric ratios of 120° on the unit circle.

A. $\sin\theta = \dfrac{\sqrt{3}}{2}$, $\cos\theta = -\dfrac{1}{2}$, $\tan\theta = -\sqrt{3}$,

$\csc\theta = \dfrac{2\sqrt{3}}{3}$, $\sec\theta = -2$, $\cot\theta = -\dfrac{\sqrt{3}}{3}$

Start by finding the reference angle for 120° on the unit circle. Knowing the angle is in Quadrant II, the reference angle is $\theta' = 180° - 120° = 60°$. You now have a 30-60-90 triangle!

Refer to the unit circle in Figure 6-2, and you find that $x = -\dfrac{1}{2}$ and $y = \dfrac{\sqrt{3}}{2}$.

Now you can easily find the trig ratios using the point-in-plane definition. Keep in mind that $r = 1$. You start with sine, cosine, and tangent:

$\sin\theta = y = \dfrac{\sqrt{3}}{2}$, $\cos\theta = x = -\dfrac{1}{2}$,

$\tan\theta = \dfrac{y}{x} = \dfrac{\sqrt{3}/2}{-1/2} = \dfrac{\sqrt{3}}{-1} = -\sqrt{3}$.

Next, do the reciprocal ratios:

$\csc\theta = \dfrac{1}{y} = \dfrac{1}{\sqrt{3}/2} = \dfrac{2}{\sqrt{3}} = \dfrac{2\sqrt{3}}{3}$, $\sec\theta = \dfrac{1}{x} =$

$\dfrac{1}{-1/2} = -2$, $\cot\theta = \dfrac{x}{y} = \dfrac{-1/2}{\sqrt{3}/2} = -\dfrac{1}{\sqrt{3}} = -\dfrac{\sqrt{3}}{3}$.

Q. What's the value of θ when $\sin\theta = \dfrac{1}{2}$ and $90° < \theta < 360°$?

A. $\theta = 150°$. On a unit circle, $\sin\theta = \dfrac{1}{2}$ when $\theta = 30°$ and $\theta = 150°$. Because you're limited to $90° < \theta < 360°$, the answer is just 150°.

15 Evaluate the six trigonometric ratios of 225°.

16 Find all values of θ satisfying $\cos\theta = \dfrac{\sqrt{3}}{2}$ and $0° < \theta < 360°$.

 17 Evaluate the six trigonometric ratios of 330°.

 18 Find all values of θ satisfying $\tan\theta = \dfrac{\sqrt{3}}{3}$ and $0° < \theta < 360°$.

Solving Trig Equations

Solving trigonometric equations is just like solving regular algebraic equations, with a twist! The twist is the trigonometric term. Instead of isolating the variable, you need to isolate the trigonometric term. From there, you can use the handy unit circle to find your solution. For a complete unit circle, refer to Figure 6-2.

TIP

Given what you already know about co-terminal angles, you know that any given trig equation may have the possibility of an infinite number of solutions. Therefore, for these examples, stick with angles that are within one positive rotation of the unit circle $0 < x < 2\pi$. But make sure that you check for multiple solutions within that unit circle!

Also, Figure 6-4 provides a handy chart of the three main trig functions and their values at the basic angles. It's a nice, quick reference for when solving equations.

Deg	0	30	45	60	90	120	135	150	180
Rad	0	$\frac{\pi}{6}$	$\frac{\pi}{4}$	$\frac{\pi}{3}$	$\frac{\pi}{2}$	$\frac{2\pi}{3}$	$\frac{3\pi}{4}$	$\frac{5\pi}{6}$	π
Sin	0	$\frac{1}{2}$	$\frac{\sqrt{2}}{2}$	$\frac{\sqrt{3}}{2}$	1	$\frac{\sqrt{3}}{2}$	$\frac{\sqrt{2}}{2}$	$\frac{1}{2}$	0
Cos	1	$\frac{\sqrt{3}}{2}$	$\frac{\sqrt{2}}{2}$	$\frac{1}{2}$	0	$-\frac{1}{2}$	$-\frac{\sqrt{2}}{2}$	$-\frac{\sqrt{3}}{2}$	-1
Tan	0	$\frac{\sqrt{3}}{3}$	1	$\sqrt{3}$	und	$-\sqrt{3}$	-1	$-\frac{\sqrt{3}}{3}$	0

Deg		210	225	240	270	300	315	330	360
Rad		$\frac{7\pi}{6}$	$\frac{5\pi}{4}$	$\frac{4\pi}{3}$	$\frac{3\pi}{2}$	$\frac{5\pi}{3}$	$\frac{7\pi}{4}$	$\frac{11\pi}{6}$	2π
Sin		$-\frac{1}{2}$	$-\frac{\sqrt{2}}{2}$	$-\frac{\sqrt{3}}{2}$	-1	$-\frac{\sqrt{3}}{2}$	$-\frac{\sqrt{2}}{2}$	$-\frac{1}{2}$	0
Cos		$-\frac{\sqrt{3}}{2}$	$-\frac{\sqrt{2}}{2}$	$-\frac{1}{2}$	0	$\frac{1}{2}$	$\frac{\sqrt{2}}{2}$	$\frac{\sqrt{3}}{2}$	1
Tan		$\frac{\sqrt{3}}{3}$	1	$\sqrt{3}$	und	$-\sqrt{3}$	-1	$-\frac{\sqrt{3}}{3}$	0

FIGURE 6-4: Some trig functions at the most popular angles.

Q. Solve for x in $2\sin x = 1$; x is in radians.

EXAMPLE

A. $x = \frac{\pi}{6}$, or $x = \frac{5\pi}{6}$. Because you already know how to solve $2y = 1$, you also know how to solve $2\sin x = 1$: It's $\sin x = \frac{1}{2}$.

The question is what to do with it from there. Well, now you need to find the angle or angles that make the equation true. Here's where that unit circle comes in handy! Remembering that $\sin\theta = y$, you can look at the unit circle to find which angles have $y = \frac{1}{2}$. The two angles are $x = \frac{\pi}{6}$ and $x = \frac{5\pi}{6}$. Or you can refer to the handy chart, since you're dealing with sine.

Q. Solve $2\cos^2 x - \cos x = 1$, giving answers in terms of degrees.

A. $x = 0°, 120°, 240°, 360°$. Observe that $\cos^2 x = (\cos x)^2$; letting $\cos x = z$ you can think of the equation as $2z^2 - z = 1$. You see that it's a simple quadratic that you need to try and factor and then solve using the zero product property: $2z^2 - z - 1 = 0$ factors into $(2z + 1)(z - 1) = 0$. Now change back to the trig function and you have $(2\cos x + 1)(\cos x - 1) = 0$. Using the zero product property, when $2\cos x + 1 = 0$, then $\cos x = -\dfrac{1}{2}$. For the other factor,

when $\cos x - 1 = 0$, then $\cos x = 1$. Now it's time to use those reference angles! Ask yourself this: When is $\cos x = -\dfrac{1}{2}$? Referring to the unit circle, you see that your reference angle θ' is 60°, and your answer falls in Quadrants II and III. Therefore, the resulting angles are in Quad II ($180° - 60° = 120°$) and Quad III ($180° + 60° = 240°$). For your second equation, $\cos x = 1$ and $x = 0°$ or 360°. Therefore, your four solutions are $x = 0°,\ 120°,\ 240°,\ 360°$.

---

19 Solve for θ in $3\tan\theta - 1 = 2$ where $0 < \theta < 360°$.

20 Solve for θ in $\sin^2\theta = \sin\theta$ where $0 < \theta < 2\pi$.

21 Solve for θ in $2\cos^2\theta - 1 = 0$ where $0 < \theta < 2\pi$.

22 Solve for θ in $4\sin^2\theta + 3 = 4$ where $0 < \theta < 2\pi$.

23 Solve for θ in $4\sin^4\theta - 7\sin^2\theta + 3 = 0$ where $0 < \theta < 360°$.

24 Solve for θ in $\tan^2\theta - \tan\theta = 0$ where $0 < \theta < 2\pi$.

Making and Measuring Arcs

If someone asks you how far an ant on the edge of a 6-inch CD travels if the CD spins a mere 120°, you probably wonder why it matters. You may even be thinking that the ant is probably messing up your CD player! But wacky math teachers love coming up with questions like that, so I'm here to help you solve them.

To calculate the measure of an *arc*, a portion of the circumference of a circle like the path that pesky ant is taking, you need to remember that arcs can be measured in two ways: as an angle and as a length. As an angle, there's nothing to calculate — the measure of the arc is simply the same as the measure of the central angle. As a length, the measure of the arc is directly proportional to the circumference of the circle. If θ is measured in degrees, r is the radius, and s is the arc length cut off by the angle θ, then the ratio of s over the circumference of the circle

is the same as the angle value in degrees over 360. This gives you the formula $s = 2\pi r\left(\dfrac{\theta}{360}\right)$.

When θ is expressed in radians, the formula becomes simply $s = \theta r$. Figure 6-5 shows the situation.

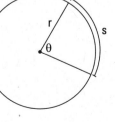

FIGURE 6-5:
Calculating arc length and the variables involved.

EXAMPLE

Q. Back to that ant! A pesky ant is on the edge of a 6-inch CD. How far does the ant travel if the CD spins 120°?

A. The ant travels approximately 6.3 inches. You can use either formula. However, to use the second formula, the angle needs to be in radians, so using the proportion $\dfrac{\theta°}{180°} = \dfrac{\theta^R}{\pi}$, insert the 120° to

get $\dfrac{120°}{180°} = \dfrac{2}{3} = \dfrac{\theta^R}{\pi}$. Multiplying by π, you have $\theta^R = \dfrac{2\pi}{3}$. The diameter is 6 inches, so the radius is 3 inches. Using the formula $s = \theta r$, you get $s = \dfrac{2\pi}{3} \cdot 3 = 2\pi$ inches, which is approximately 6.3 inches.

---

25 Find the length of an arc in a circle with a radius of 4 feet if the central angle is $\dfrac{\pi}{6}$.

26 Find the length of an arc in a circle with a diameter of 16 centimeters if the central angle is $\dfrac{7\pi}{4}$.

27 Find the length of an arc in a circle with a radius of 18 feet if the angle is 210°.

28 Find the radius of a circle in which an angle of 2 radians cuts an arc with a length of 42 inches.

Answers to Problems on Basic Trig and the Unit Circle

This section contains the answers for the practice problems presented in this chapter, plus explanations on how to arrive at those answers.

(1) Find $\cos A$ in $\triangle ABC$. The answer is $\cos A = \dfrac{5\sqrt{41}}{41}$.

Because $\cos\theta = \dfrac{\text{adj}}{\text{hyp}}$, you need to find the adjacent side using the Pythagorean Theorem:

$x^2 + 4^2 = \left(\sqrt{41}\right)^2$, $x^2 + 16 = 41$, $x^2 = 25$, $x = 5$. Therefore, $\cos A = \dfrac{5}{\sqrt{41}} = \dfrac{5\sqrt{41}}{41}$.

(2) Solve $\triangle DEF$. The answer is $\angle E = 65°$, side $e \approx 23.6\text{cm}$, and side $f \approx 26$ cm.

First, because you know $\angle D$ and $\angle F$, you can find $\angle E$ by subtracting the sum of $\angle D$ and $\angle F$ from $180°$: $180° - (25° + 90°) = 65°$. To find side e, you can use $\tan 65° = \dfrac{e}{11}$. Multiplying both sides by 11, you get $e = 11 \cdot \tan 65°$, which is approximately 23.6. To find side f, you can use $\sin 25° = \dfrac{11}{f}$. Multiply both sides by f and divide by the sine to get: $f = \dfrac{11}{\sin 25°}$, which is approximately 26 cm.

(3) Find $\angle Q$ in $\triangle QRS$ (round to the nearest tenth). The answer is approximately $22.6°$.

Because you have the adjacent side to $\angle Q$ and the hypotenuse, you use cosine: $\cos Q = \dfrac{12}{13}$. To solve, take the inverse cosine of each side: $Q = \cos^{-1}\left(\dfrac{12}{13}\right)$, which is approximately $22.6°$.

(4) $\angle R$ in $\triangle QRS$ from Question 3. The answer is $\sin R = \dfrac{12}{13}$, $\cos R = \dfrac{5}{13}$, $\tan R = \dfrac{12}{5}$, $\csc R = \dfrac{13}{12}$, $\sec R = \dfrac{13}{5}$, $\cot R = \dfrac{5}{12}$.

Start by using the Pythagorean Theorem to find the third side: $12^2 + q^2 = 13^2$, $q^2 = 13^2 - 12^2 = 25$, $q = 5$. Then, plug the sides into the trigonometric ratios:

$\sin R = \dfrac{12}{13}$, $\cos R = \dfrac{5}{13}$, $\tan R = \dfrac{12}{5}$, $\csc R = \dfrac{13}{12}$, $\sec R = \dfrac{13}{5}$, $\cot R = \dfrac{5}{12}$

(5) Romero wants to deliver a rose to his girlfriend, Jules, who is sitting on her balcony 24 feet above the street. If Romero has a 28-foot ladder, at what angle must he place the bottom of the ladder to reach his love, Jules? The answer is $59°$.

To solve, draw a picture:

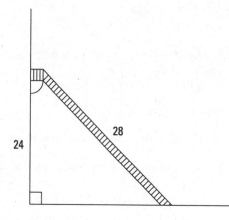

With the picture, you can see that you have the opposite side, 24 feet, opposite the angle you want and the hypotenuse, 28 feet. Therefore, you can use the sine ratio to solve: $\sin\theta = \dfrac{24}{28}$.
To solve for the angle, use inverse sine: $\theta = \sin^{-1}\left(\dfrac{24}{28}\right)$, which is approximately 59°.

6. Sam needs to cross a river. He spies a bridge directly ahead of him. Looking across the river, he sees that he's 27° below the bridge from the other side. How far must he walk on his side of the river to reach the bridge if the bridge length is 40 feet? The answer is 79 more feet.

First, consider the picture. You'll find it with the original problem, earlier in this chapter.

Considering that you have the opposite side from the angle, 40 feet, and you're looking for the adjacent side, you can use the tangent ratio: $\tan 27° = \dfrac{40}{x}$ Multiplying both sides by x and then dividing by the tangent, you get $x = \dfrac{40}{\tan 27°}$, which equals approximately 79 feet.

7. Paul, a 6-foot-tall man, is holding the end of a kite string 5 feet above the ground. The string of the kite is 75-feet long with 35° of elevation. Paulette, Paul's 5-foot-tall daughter, is directly underneath the kite. How far above Paulette's head is the kite? The answer is about 43 feet.

Begin by (you guessed it!) drawing a picture:

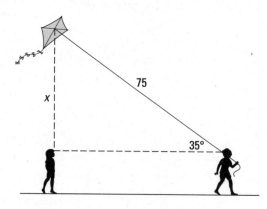

Because Paul is holding the end of the kite string at the same height as Paulette's head, you only need to consider the string of the kite, which forms the hypotenuse of the triangle and the angle. Because you're looking for the opposite side from the angle and you have the hypotenuse, use the sine ratio to solve: $\sin 35° = \dfrac{x}{75}$. Multiplying both sides by 75, you get $x = 75 \cdot \sin 35°$, which is approximately 43 feet.

8 To hold up a 100-foot pole, two guide wires are put up on opposite sides of the pole. One wire makes a 36° angle with the ground and the other makes a 47° angle with the ground. How far apart are the bases of the wires? The answer is about 231 feet apart.

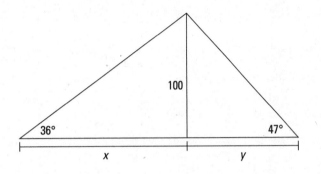

Using your picture, you can set up two tangent ratios: $\tan 36° = \dfrac{100}{x}$ and $\tan 47° = \dfrac{100}{y}$. Solving for x and y, respectively: $x = \dfrac{100}{\tan 36°}$ and $y = \dfrac{100}{\tan 47°}$. Therefore, x is approximately 137.6 feet and y, is approximately 93.3. Add these together to get that the total distance apart is about 231 feet.

9 Find three co-terminal angles of $\dfrac{\pi}{5}$. Although there are multiple answers, three possible answers are $\dfrac{11\pi}{5}$, $-\dfrac{9\pi}{5}$, and $\dfrac{21\pi}{5}$.

Simply add or subtract multiples of 2π: $\dfrac{\pi}{5} + 2\pi = \dfrac{11\pi}{5}$; $\dfrac{\pi}{5} - 2\pi = -\dfrac{9\pi}{5}$; $\dfrac{\pi}{5} + 2 \cdot 2\pi = \dfrac{11\pi}{5}$.

10 Find two positive co-terminal angles of −775°. Two possible answers are 305° and 665°.

Here, just add multiples of 360° to −775° until you get two positive co-terminal angles: $-775° + 3 \cdot 360° = 305°$; $-775° + 4 \cdot 360° = 665°$.

11 Evaluate the six trigonometric ratios of the point (3, 4). The answers are $\sin\theta = \dfrac{4}{5}$, $\cos\theta = \dfrac{3}{5}$, $\tan\theta = \dfrac{4}{3}$, $\csc\theta = \dfrac{5}{4}$, $\sec\theta = \dfrac{5}{3}$, $\cot\theta = \dfrac{3}{4}$.

First, find the radius using the Pythagorean Theorem: $3^2 + 4^2 = r^2$, $r^2 = 25$, $r = 5$. Using this and $x = 3$, $y = 4$, plug into the trigonometric ratios: $\sin\theta = \dfrac{4}{5}$, $\cos\theta = \dfrac{3}{5}$, $\tan\theta = \dfrac{4}{3}$, $\csc\theta = \dfrac{5}{4}$, $\sec\theta = \dfrac{5}{3}$, $\cot\theta = \dfrac{3}{4}$.

(12) Evaluate the six trigonometric ratios of the point $(-5, -7)$. The answers are
$\sin\theta = -\dfrac{7\sqrt{74}}{74}, \cos\theta = -\dfrac{5\sqrt{74}}{74}, \tan\theta = \dfrac{7}{5}$, and $\csc\theta = -\dfrac{\sqrt{74}}{7}, \sec\theta = -\dfrac{\sqrt{74}}{5}, \cot\theta = \dfrac{5}{7}$.

Find your radius: $(-5)^2 + (-7)^2 = r^2, r^2 = 74, r = \sqrt{74}$. Using this and the point (x, y), plug into the trig ratios and rationalize if necessary: $\sin\theta = -\dfrac{7\sqrt{74}}{74}, \cos\theta = -\dfrac{5\sqrt{74}}{74}, \tan\theta = \dfrac{7}{5}$ and $\csc\theta = -\dfrac{\sqrt{74}}{7}, \sec\theta = -\dfrac{\sqrt{74}}{5}, \cot\theta = \dfrac{5}{7}$.

(13) Evaluate the six trigonometric ratios of the point $\left(-2, 2\sqrt{3}\right)$. The answers are
$\sin\theta = \dfrac{\sqrt{3}}{2}, \cos\theta = -\dfrac{1}{2}, \tan\theta = -\sqrt{3}$, and $\csc\theta = \dfrac{2\sqrt{3}}{3}, \sec\theta = -2, \cot\theta = -\dfrac{\sqrt{3}}{3}$.

Begin by finding your radius: $r^2 = (-2)^2 + \left(2\sqrt{3}\right)^2 = 4 + 12 = 16, r = 4$. Now plug into your trig ratios and rationalize if necessary: $\sin\theta = \dfrac{2\sqrt{3}}{4} = \dfrac{\sqrt{3}}{2}, \cos\theta = \dfrac{-2}{4} = -\dfrac{1}{2}, \tan\theta = \dfrac{2\sqrt{3}}{-2} = -\sqrt{3}$,
$\csc\theta = \dfrac{4}{2\sqrt{3}} = \dfrac{2\sqrt{3}}{3}, \sec\theta = \dfrac{4}{-2} = -2, \cot\theta = \dfrac{-2}{2\sqrt{3}} = -\dfrac{\sqrt{3}}{3}$.

(14) Evaluate the six trigonometric ratios of the point $\left(6, -3\sqrt{5}\right)$. The answers are
$\sin\theta = -\dfrac{\sqrt{5}}{3}, \cos\theta = \dfrac{2}{3}, \tan\theta = -\dfrac{\sqrt{5}}{2}, \csc\theta = -\dfrac{3\sqrt{5}}{5}, \sec\theta = \dfrac{3}{2}, \cot\theta = -\dfrac{2\sqrt{5}}{5}$.

Start by finding the radius: $r^2 = (6)^2 + \left(-3\sqrt{5}\right)^2 = 36 + 45 = 81, r = 9$. Plug into your trig ratios and rationalize if necessary: $\sin\theta = \dfrac{-3\sqrt{5}}{9} = -\dfrac{\sqrt{5}}{3}, \cos\theta = \dfrac{6}{9} = \dfrac{2}{3}, \tan\theta = \dfrac{-3\sqrt{5}}{6} - \dfrac{\sqrt{5}}{2}$,
$\csc\theta = \dfrac{9}{-3\sqrt{5}} = -\dfrac{3\sqrt{5}}{5}, \sec\theta = \dfrac{9}{6} = \dfrac{3}{2}, \cot\theta = \dfrac{6}{-3\sqrt{5}} = -\dfrac{2\sqrt{5}}{5}$.

(15) Evaluate the six trigonometric ratios of $225°$. The answers are $\sin\theta = -\dfrac{\sqrt{2}}{2}, \cos\theta = -\dfrac{\sqrt{2}}{2}$, $\tan\theta = 1$, and $\csc\theta = -\sqrt{2}, \sec\theta = -\sqrt{2}, \cot\theta = 1$.

Using reference angles, you can see that you're dealing with a 45-45-90 triangle $(225° - 180°)$. Therefore, $x = -\dfrac{\sqrt{2}}{2}$ and $y = -\dfrac{\sqrt{2}}{2}$. Now, by using the point-in-plane definition, you can find the six trigonometric ratios: $\sin\theta = y = -\dfrac{\sqrt{2}}{2}, \cos\theta = x = -\dfrac{\sqrt{2}}{2}, \tan\theta = \dfrac{y}{x} = 1$,
$\csc\theta = \dfrac{1}{y} = -\sqrt{2}, \sec\theta = \dfrac{1}{x} = -\sqrt{2}, \cot\theta = \dfrac{x}{y} = 1$. You also have the first three values available in the chart and can just find the reciprocals for the reciprocal functions.

(16) Find all values of θ satisfying $\cos\theta = \dfrac{\sqrt{3}}{2}$ and $0° < \theta < 360°$. The answer is $30°$ and $330°$.

Looking at the special right triangles, you can see that $\cos\theta = \dfrac{\sqrt{3}}{2}$ when $\theta = 30°$. Because cosine is equal to the x value on the unit circle and x is positive in Quadrants I and IV, the answer is $30°$ and $330°$.

(17) Evaluate the six trigonometric ratios of 330°. The answers are $\sin\theta = -\dfrac{1}{2}$, $\cos\theta = \dfrac{\sqrt{3}}{2}$, $\tan\theta = -\dfrac{\sqrt{3}}{3}$, and $\csc\theta = -2$, $\sec\theta = \dfrac{2\sqrt{3}}{3}$, $\cot\theta = -\sqrt{3}$.

Considering that 330° is in Quadrant IV, using reference angles (360° − 330°), you find that you're dealing with a 30-60-90 triangle. Using the point-in-plane definition, you get

$\sin\theta = y = -\dfrac{1}{2}$, $\cos\theta = x = \dfrac{\sqrt{3}}{2}$, $\tan\theta = \dfrac{y}{x} = -\dfrac{\sqrt{3}}{3}$, $\csc\theta = \dfrac{1}{y} = -2$, $\sec\theta = \dfrac{1}{x} = \dfrac{2\sqrt{3}}{3}$, $\cot\theta = \dfrac{x}{y} = -\sqrt{3}$.

You also have the first three values available in the chart and can just find the reciprocals for the reciprocal functions.

(18) Find all values of θ satisfying $\tan\theta = \dfrac{\sqrt{3}}{3}$ and $0° < \theta < 360°$. The answer is $\theta = 30°$ and $\theta = 210°$.

To solve this, use the unit circle, the chart of function values, or the inverse tangent to find that $\theta = 30°$. Using the unit circle, look for when y is $\dfrac{1}{2}$ and x is $\dfrac{\sqrt{3}}{2}$, giving you $\dfrac{y}{x}$ with the target value. Because the tangent value is positive, both sine and cosine must be the same sign, which occurs in Quadrants I and III. To get the Q III angle, just add 30° + 180° = 210°. Therefore, $\theta = 30°$ and $\theta = 210°$.

(19) Solve for θ in $3\tan\theta - 1 = 2$ where $0 < \theta < 360°$. The answer $\theta = 45°$ and $\theta = 225°$.

Begin by using algebra to isolate $\tan\theta$: $3\tan\theta - 1 = 2$, $3\tan\theta = 3$, and $\tan\theta = 1$. Using the unit circle or inverse tangent, you find when $\tan\theta$ is to equal 1 or when the x and y values are the same. This occurs when $\theta = 45°$. Because the answer is positive, both sine and cosine must be the same sign, which occurs in Quadrants I and III. Therefore, using reference angles, for Quadrant I, $\theta = 45°$, and for Quadrant III, 45° + 180° = 225°.

(20) Solve for θ in $\sin^2\theta = \sin\theta$ where $0 < \theta < 2\pi$. The answer is $\theta = 0, \dfrac{\pi}{2}, \pi, 2\pi$.

To solve, think of $\sin^2\theta = \sin\theta$ as $x^2 = x$, which can be solved by bringing both terms to the same side and factoring: $x^2 - x = 0$, $x(x-1) = 0$. Similarly, $\sin^2\theta = \sin\theta$, $\sin^2\theta - \sin\theta = 0$, $\sin\theta(\sin\theta - 1) = 0$. Therefore, $\sin\theta = 0$, or $\sin\theta - 1 = 0$, which means $\sin\theta = 1$. Knowing that $\sin\theta = y$ on the unit circle, $\sin\theta = 0$ at 0, π, and 2π; also, $\sin\theta = 1$ at $\dfrac{\pi}{2}$, you have your answers!

(21) Solve for θ in $2\cos^2\theta - 1 = 0$ where $0 < \theta < 2\pi$. The answer is $\theta = \dfrac{\pi}{4}, \dfrac{3\pi}{4}, \dfrac{5\pi}{4}, \dfrac{7\pi}{4}$.

First, isolate the cosine term using algebra: $2\cos^2\theta - 1 = 0$, $2\cos^2\theta = 1$, $\cos^2\theta = \dfrac{1}{2}$. Now, take the square root of each side: $\cos\theta = \pm\sqrt{\dfrac{1}{2}} = \pm\dfrac{\sqrt{2}}{2}$. This occurs at four angles on the unit circle: $\dfrac{\pi}{4}, \dfrac{3\pi}{4}, \dfrac{5\pi}{4}$, and $\dfrac{7\pi}{4}$.

(22) Solve for θ in $4\sin^2\theta + 3 = 4$ where $0 < \theta < 2\pi$. The answer is $\theta = \dfrac{\pi}{6}, \dfrac{5\pi}{6}, \dfrac{7\pi}{6}, \dfrac{11\pi}{6}$.

Begin by using algebra to isolate the sine term: $4\sin^2\theta + 3 = 4$, $4\sin^2\theta = 1$, $\sin^2\theta = \dfrac{1}{4}$. Taking the square root of each side, you get: $\sin\theta = \pm\sqrt{\dfrac{1}{4}} = \pm\dfrac{1}{2}$. This tells you that $\theta = \dfrac{\pi}{6}, \dfrac{5\pi}{6}, \dfrac{7\pi}{6}, \dfrac{11\pi}{6}$.

(23) Solve for θ in $4\sin^4\theta - 7\sin^2\theta + 3 = 0$ where $0 < \theta < 360°$. The answer is $\theta = 60°, 90°, 120°, 240°,$ $240°, 270°, 300°$.

Start by thinking of $4\sin^4\theta - 7\sin^2\theta + 3 = 0$ as $4x^4 - 7x^2 + 3 = 0$, which factors into $(4x^2 - 3)(x^2 - 1) = 0$. Similarly, $4\sin^4\theta - 7\sin^2\theta + 3 = 0$ factors into $(4\sin^2\theta - 3)(\sin^2\theta - 1) = 0$. Set each factor equal to zero and take the square root of each side to find $\sin\theta$: $4\sin^2\theta - 3 = 0$, $\sin^2\theta = \dfrac{3}{4}$ $\sin\theta = \pm\sqrt{\dfrac{3}{4}} = \pm\dfrac{\sqrt{3}}{2}$. The angles with either of these values for their sine are $\theta = 60°, 120°, 240°, 300°$. Then, when $\sin^2\theta - 1 = 0$, $\sin^2\theta = 1$, $\sin\theta = \pm1$. This gives you $\theta = 90°, 270°$.

(24) Solve for θ in $\tan^2\theta - \tan\theta = 0$ where $0 < \theta < 2\pi$. The answer is $\theta = 0, \pi, 2\pi, \dfrac{\pi}{4}, \dfrac{5\pi}{4}$.

Notice that this problem is similar to Question 20. You can factor the same way: $\tan^2\theta - \tan\theta = 0$, $\tan\theta(\tan\theta - 1) = 0$. Set each factor to zero: $\tan\theta = 0$ or $\tan\theta - 1 = 0$, so $\tan\theta = 1$. These occur at $\theta = 0, \pi, 2\pi, \dfrac{\pi}{4}, \dfrac{5\pi}{4}$.

(25) Find the length of an arc in a circle with a radius of 4 feet if the central angle is $\dfrac{\pi}{6}$. The answer is $s = \dfrac{2\pi}{3}$ feet, or approximately 2.1 feet.

Use the formula $s = \theta \cdot r$, giving you $s = \dfrac{\pi}{6} \cdot 4 = \dfrac{2\pi}{3}$.

(26) Find the length of an arc in a circle with a diameter of 16 centimeters if the central angle is $\dfrac{7\pi}{4}$. The answer is $s = 14\pi$ cm, which is approximately 44 cm.

Start by finding the radius by dividing the diameter by two: 8 cm. Next, plug into the arc length formula: $s = \dfrac{7\pi}{4} \cdot 8 = 14\pi \approx 44$.

(27) Find the length of an arc in a circle with a radius of 18 feet if the angle is $210°$. The answer is $s = 21\pi$ feet, which is approximately 66 feet.

Begin by changing degrees to radians by using the proportion $\dfrac{\theta°}{180°} = \dfrac{\theta^R}{\pi}$, which gives you $\dfrac{210°}{180°} = \dfrac{\theta^R}{\pi}, \dfrac{7}{6} = \dfrac{\theta^R}{\pi}$, or $\dfrac{7\pi}{6} = \theta^R$. Now, plug the radius and angle into the arc length formula: $s = \dfrac{7\pi}{6} \cdot 18 = 21\pi$ feet, which is approximately 66 feet.

(28) Find the radius of a circle in which an angle of 2 radians cuts an arc of length 42 inches. The answer is 21 inches.

Just plug what you have into the arc length formula, $s = \theta \cdot r$: $42 = 2 \cdot r$. Dividing both sides by 2, you have $r = 21$.

Chapter **7**

Graphing and Transforming Trig Functions

G raphing a trig function is similar to graphing any other function. You simply insert values into the input to find the output. In this case, the input is typically θ and the output is typically *y*. And, just like graphing any other function, knowing the *parent trig graph* — the most basic, unaltered graph — makes the task of graphing more complex graphs much easier. In this chapter, you see the parent graph of each trig function and some of its transformations.

Getting a Grip on Periodic Graphs

Periodic graphs are like other graphs of functions that keep going, and going, and going. But these graphs have a really special property: They keep repeating and repeating the same set of values over and over. Just remember that trig functions are periodic graphs, and the steps to graphing them will be easy! Because they repeat their values over and over again, you just need to figure out one period (or cycle), and then you can repeat it as many times as you like.

REMEMBER

The key to graphing trig functions is to graph just one period. You start by graphing the parent graph, and then perform any necessary transformations (just like you do in Chapter 3 for other types of graphs). As with other graphs, the same transformations can be applied to trig graphs:

>> For trig functions, vertical stretches and flattening are achieved by simply multiplying the parent function by a constant. For example, $f(\theta) = 2\sin\theta$ is the same as the parent graph, only its wave goes up to a value of 2 and down to -2. Multiplying the parent graph by a negative constant simply flips the graph upside down, or reflects it over the x-axis.

>> Horizontal stretches and compressions occur by changing the period of the graph. For sine and cosine parent graphs, the period is 2π. The same is true for cosecant and secant graphs. For tangent and cotangent graphs, the period is π. Multiplying the angle in the function by a constant transforms the period. For example, $f(\theta) = \cos 2\theta$ results in a graph that repeats itself twice in the amount of space the parent graph would occupy.

Vertical and horizontal translations shift the parent graph up, down, left, or right:

>> Just as you see in Chapter 3, vertical and horizontal shifts just change the location of the parent graph: up, down, left, or right.

>> The general equation for these shifts, for example, is $f(\theta) = \sin(\theta - h) + k$, where h represents the horizontal shift left or right and k represents the vertical shift up or down.

>> To find the horizontal shift of a function, simply set the inside parentheses equal to 0. For example, the horizontal shift for $\sin(\theta + 3)$ is -3 because $\theta + 3 = 0$, or $\theta = -3$.

TIP

When you put the transformations together, it's best to follow this simple order:

1. **Adjust the amplitude of the function, if applicable.**

2. **Adjust the period of the function, if applicable.**

3. **Shift the graph horizontally, if applicable.**

4. **Shift the graph vertically, if applicable.**

Parent Graphs and Transformations: Sine and Cosine

Sine and cosine graphs look like waves. These waves, or *sinusoids* in math speak, keep going and going without change. To graph these sinusoids, you need to start by checking out the parent graphs (see Figure 7-1).

Notice that the periods of both the sine and cosine graphs are the same: 2π. Similarly, they both have an amplitude (or height) of 1. You use this information for your transformations.

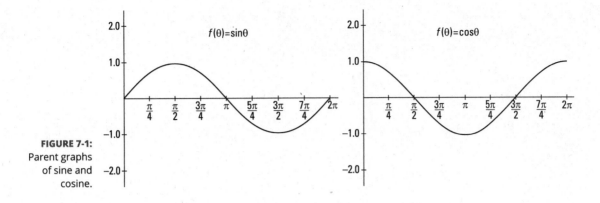

FIGURE 7-1:
Parent graphs
of sine and
cosine.

Putting together all the transformation information into one equation, you get

$$f(\theta) = a \cdot \sin\left[p(\theta - h)\right] + k$$

$$f(\theta) = a \cdot \cos\left[p(\theta - h)\right] + k$$

where a is the amplitude, h is the horizontal shift, k is the vertical shift, and you divide 2π by p to get the period.

Q. Graph $f(\theta) = 2\sin\theta + 3$.

EXAMPLE **A.** Starting with amplitude, you can see that $a = 2$, so your amplitude is 2. The period is 2π because the period doesn't change from the parent equation — there's no multiplier on θ. The vertical shift is positive 3 because $k = 3$. The graph is shown in the following figure.

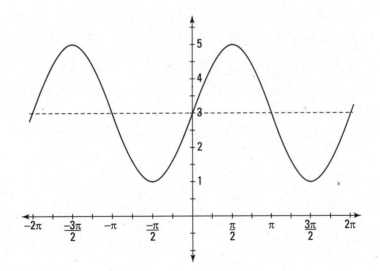

1 Graph $f(\theta) = -\dfrac{1}{2}\cos\theta$.

2 Graph $f(\theta) = \cos\dfrac{1}{2}\theta$.

3 Graph $f(\theta) = \sin\left(\theta + \dfrac{\pi}{4}\right)$.

4 Graph $f(\theta) = \cos\dfrac{1}{3}\theta + 2$.

5 Name the amplitude, period, horizontal shift, and vertical shift of $f(\theta) = 3\sin\left(2\theta + \dfrac{\pi}{2}\right) - 1$.

6 Graph $f(\theta) = 3\sin\left(2\theta + \dfrac{\pi}{2}\right) - 1$.

Tangent and Cotangent: More Family Members

Tangent and cotangent are both periodic, but they're not wavelike like sine and cosine. Instead, they have vertical asymptotes that break up their graphs. As discussed in Chapter 3, a *vertical asymptote* can occur where the function is undefined. Because tangent is defined $\tan\theta = \dfrac{\sin\theta}{\cos\theta}$, and cotangent is defined $\cot\theta = \dfrac{\cos\theta}{\sin\theta}$ are ratios, they both have values that are undefined where their denominators are equal to 0. For tangent, this occurs when $\theta = \ldots, -\dfrac{5\pi}{2}, -\dfrac{3\pi}{2}, -\dfrac{\pi}{2}, \dfrac{\pi}{2}, \dfrac{3\pi}{2}, \dfrac{5\pi}{2}, \ldots$. For cotangent, this occurs when $\theta = \ldots, -2\pi, -\pi, 0, \pi, 2\pi, \ldots$ Therefore, these are the locations of their asymptotes (see Figure 7-2).

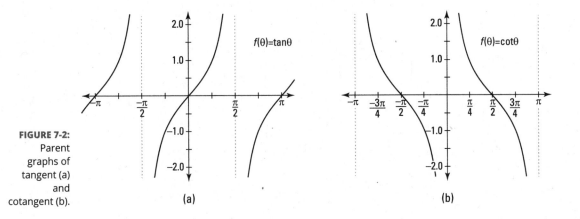

(a)

(b)

FIGURE 7-2: Parent graphs of tangent (a) and cotangent (b).

Notice that the periods of both the tangent and cotangent graphs are the same: π. The x-intercepts for tangent are $0, \pi$, and $-\pi$. For cotangent, the x-intercepts are $\dfrac{\pi}{2}$ and $-\dfrac{\pi}{2}$. Using this information, you can make your transformations.

TECHNICAL STUFF

Putting together all the information from earlier in this chapter into one equation, you get

$$f(\theta) = a \cdot \tan\left[p(\theta - h)\right] + k$$

$$f(\theta) = a \cdot \cot\left[p(\theta - h)\right] + k$$

where a is the vertical change (no amplitude with tangent and cotangent), h is the horizontal shift, k is the vertical shift, and you divide π by p to get the period.

Q. Graph $f(\theta) = \dfrac{1}{2}\tan 2\theta$.

EXAMPLE **A.** Starting with the vertical change, you can see that it's a flattening, because the multiplier is $\dfrac{1}{2}$. Next, find the period by dividing π by 2, which is $\dfrac{\pi}{2}$. Because there are no horizontal or vertical shifts, you're ready to graph:

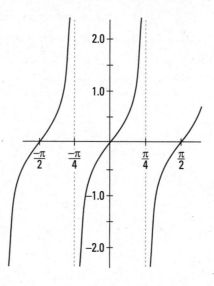

7 Graph $f(\theta) = \cot\dfrac{1}{2}\theta$.

8 Graph $f(\theta) = \tan\theta + 2$.

9 Graph $f(\theta) = \dfrac{1}{3}\cot\theta$.

10 Graph $f(\theta) = \tan\left(\theta - \dfrac{\pi}{2}\right)$.

11 Name the vertical stretch or flattening, period, horizontal shift, and vertical shift of $f(\theta) = 2\tan\left(\theta + \dfrac{\pi}{4}\right) - 1$.

12 Graph $f(\theta) = 2\tan\left(\theta + \dfrac{\pi}{4}\right) - 1$.

Generations: Secant and Cosecant

To graph cosecant and secant, it's important to remember that they're the reciprocals of sine and cosine, respectively: $\csc\theta = \dfrac{1}{\sin\theta}$ and $\sec\theta = \dfrac{1}{\cos\theta}$. Using this fact, the easiest way to graph cosecant or secant is to start by graphing sine or cosine — the graphs of the reciprocals are easily found from there.

Given that cosecant and secant are reciprocal functions of sine and cosine, respectively, as sine and cosine approach zero, their reciprocals get larger and larger without limit. So, yep, you guessed it — cosecant and secant graphs have asymptotes. These occur wherever their reciprocal functions (sine or cosine) have a value of 0. To graph, follow these easy steps:

1. **Graph the sine graph with any transformations to graph a cosecant graph, or graph the cosine graph with transformations to get the secant graph.**

2. **Draw vertical asymptotes where the sine or cosine functions are equal to 0.**

3. **Sketch the reciprocal graph of sine or cosine between each pair of asymptotes.**

    For example, if the sine graph gets bigger, the cosecant graph gets smaller.

    The parent graphs of cosecant and secant are shown in Figure 7-3. The parent sine and cosine graphs are also shown so you can see where they came from.

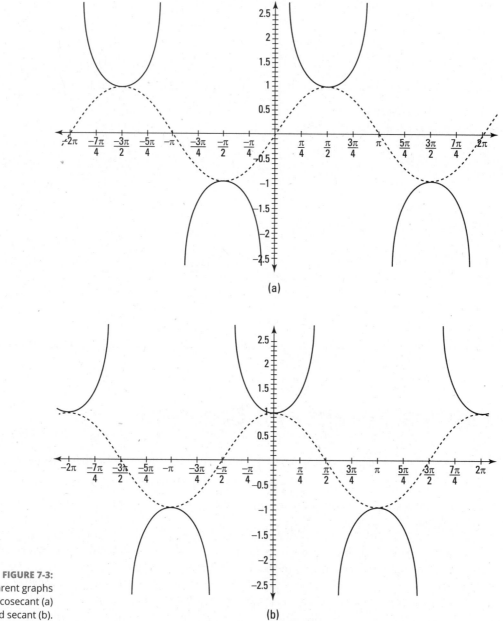

(a)

(b)

FIGURE 7-3:
Parent graphs
of cosecant (a)
and secant (b).

Q. Graph $f(\theta) = \csc\theta + 1$.

EXAMPLE **A.** Here, there's only a vertical shift of 1, which means you shift the parent graph up one (see the following figure).

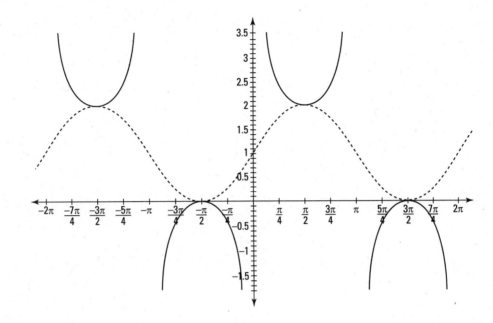

---

13 Graph $f(\theta) = -\csc\theta - 1$.

14 Graph $f(\theta) = \sec 2\theta + 1$.

15 Name the asymptotes for one period and the horizontal and vertical shifts of

$$f(\theta) = \frac{1}{4}\csc\left(\theta - \frac{\pi}{2}\right) - 1$$

16 Graph $f(\theta) = \frac{1}{4}\csc\left(\theta - \frac{\pi}{2}\right) - 1$

17 Name the period, horizontal shift, and vertical shift of $f(\theta) = 2\sec\left(\frac{1}{2}\theta + \frac{\pi}{4}\right) + 1$

18 Graph $f(\theta) = 2\sec\left(\frac{1}{2}\theta + \frac{\pi}{4}\right) + 1$

Answers to Problems on Graphing and Transforming Trig Functions

Following are the answers to problems dealing with trig functions. You also find details on getting the answers.

(1) Graph $f(\theta) = -\frac{1}{2}\cos\theta$. See the graph for the answer.

Because the cosine function is multiplied by $-\frac{1}{2}$, the graph is inverted with an amplitude of $\frac{1}{2}$. The period doesn't change, and there are no shifts.

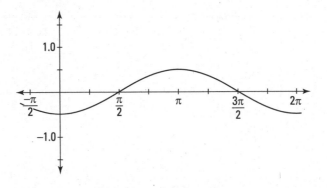

(2) Graph $f(\theta) = \cos\frac{1}{2}\theta$. See the graph for the answer.

Here, the amplitude doesn't change, but the period does. The new period is found by dividing 2π by $\frac{1}{2}$, which is 4π. There are no vertical or horizontal shifts.

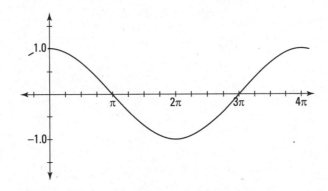

(3) Graph $f(\theta) = \sin\left(\theta + \dfrac{\pi}{4}\right)$. See the graph for the answer.

This graph has a horizontal shift. To find it, set what's inside the parentheses to the starting value of the parent graph: $\theta + \dfrac{\pi}{4} = 0$, so $\theta = -\dfrac{\pi}{4}$. This moves the graph to the left by $\dfrac{\pi}{4}$. There are no other changes from the parent graph.

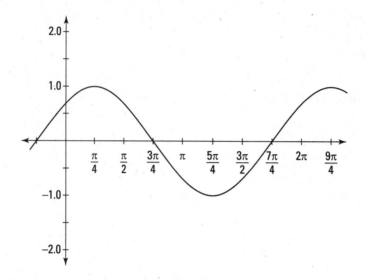

(4) Graph $f(\theta) = \cos\dfrac{1}{3}\theta + 2$. See the graph for the answer.

This has a change in period, which can be found by dividing 2π by $\dfrac{1}{3}$ to get 6π. There's also a vertical shift of 2.

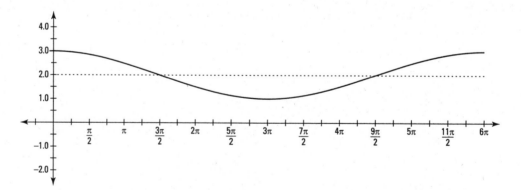

(5) Name the amplitude, period, horizontal shift, and vertical shift of $f(\theta) = 3\sin\left(2\theta + \dfrac{\pi}{2}\right) - 1$.

First, rewrite the function as $f(\theta) = 3\sin 2\left(\theta + \dfrac{\pi}{4}\right) - 1$ to make it easier to find the different changes. The amplitude is 3, the period is π, the horizontal shift is $-\dfrac{\pi}{4}$, and the vertical shift is -1.

The only calculation you need to do is to find the period. Here, you divide 2π by 2 to get π. From the equation, you can see that the amplitude is 3 and the vertical shift is -1. By setting $\theta + \dfrac{\pi}{4} = 0$, you get $\theta = -\dfrac{\pi}{4}$ or a shift to the left by $\dfrac{\pi}{4}$.

6 Graph $f(\theta) = 3\sin\left(2\theta + \dfrac{\pi}{2}\right) - 1$. See the graph for the answer.

Using the information you gather in Question 5, this graph comes together quickly.

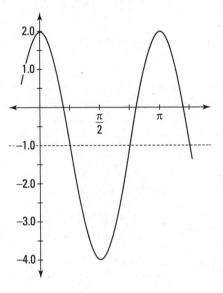

7 Graph $f(\theta) = \cot\dfrac{1}{2}\theta$. See the graph for the answer.

For this cotangent graph, the period has a change, which can be found by dividing π by $\dfrac{1}{2}$ (you get 2π). There are no other changes to the parent cotangent graph.

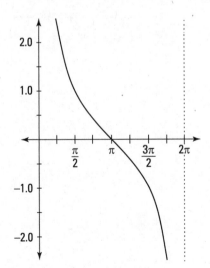

8 Graph $f(\theta) = \tan\theta + 2$. See the graph for the answer.

Here, there's only a vertical shift of 2. The period is the same as the parent graph.

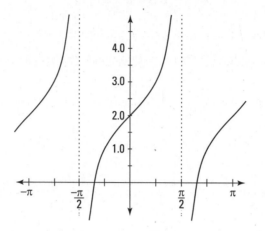

9 Graph $f(\theta) = \frac{1}{3}\cot\theta$. See the graph for the answer.

This graph flattens with the multiplication by $\frac{1}{3}$.

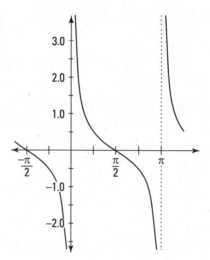

(10) Graph $f(\theta) = \tan\left(\theta - \dfrac{\pi}{2}\right)$. See the graph for the answer.

This tangent graph has a horizontal shift of $\dfrac{\pi}{2}$ to the right. There are no other changes to the parent graph.

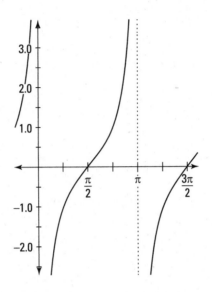

(11) Name the vertical stretch or flattening, period, horizontal shift, and vertical shift of $f(\theta) = 2\tan\left(\theta + \dfrac{\pi}{4}\right) - 1$. The vertical stretch is 2, the period is π, the horizontal shift is $-\dfrac{\pi}{4}$, and the vertical shift is -1.

The only calculation here is to find the horizontal shift by setting $\theta + \dfrac{\pi}{4} = 0$, and you get $\theta = -\dfrac{\pi}{4}$ or a shift to the left by $\dfrac{\pi}{4}$. You get the rest of the info straight from the equation.

(12) Graph $f(\theta) = 2\tan\left(\theta + \dfrac{\pi}{4}\right) - 1$. See the graph for the answer.

Using the information you gather in Question 11, you get the following graph:

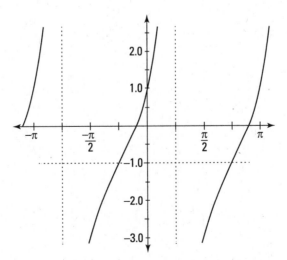

13 Graph $f(\theta) = -\csc\theta - 1$. See the graph for the answer.

Start by lightly graphing the sine curve to find the asymptotes (where the curve crosses the x-axis). Then sketch the reciprocal function. In this case, the curve is flipped over the axis and dropped by one unit.

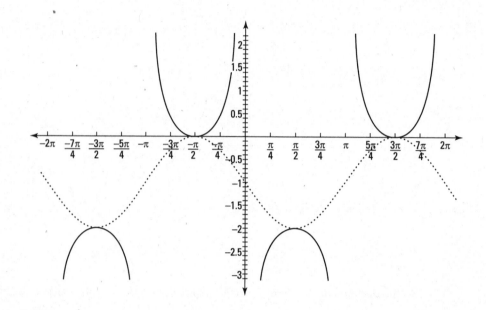

14 Graph $f(\theta) = \sec 2\theta + 1$. See the graph for the answer.

Start by lightly graphing the sine curve to find the asymptotes (where the curve crosses the x-axis). Then sketch the reciprocal function. This has a change in period, which you find by dividing 2π by 2 to get π. It also has a vertical shift of 1. Then you draw the asymptotes and sketch the secant graph.

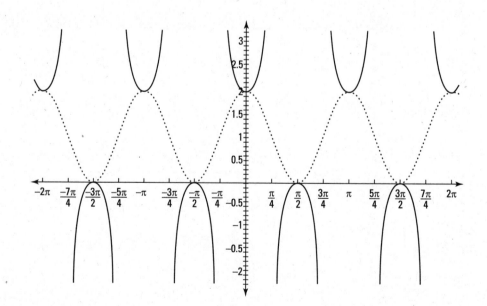

15 Name the asymptotes from 0 to 2π, and the horizontal and vertical shifts of $f(\theta) = \dfrac{1}{4}\csc\left(\theta - \dfrac{\pi}{2}\right) - 1$.

The asymptotes are at $\dfrac{\pi}{2}$ and $\dfrac{3\pi}{2}$. The horizontal shift is $\dfrac{\pi}{2}$ to the right, and the vertical shift is -1.

To find the asymptotes, you need to first look for any shifts or changes in period that would affect the parent graph. The shifts are evident from the equation, where the horizontal shift is $\dfrac{\pi}{2}$ and the vertical shift is -1. This makes the zeros of the reciprocal sine graph at $\dfrac{\pi}{2}$ and $\dfrac{3\pi}{2}$. This, then, is where the asymptotes will be.

16 Graph $f(\theta) = \dfrac{1}{4}\csc\left(\theta - \dfrac{\pi}{2}\right) - 1$. See the graph for the answer.

Using the information from Question 15, sketch the graph.

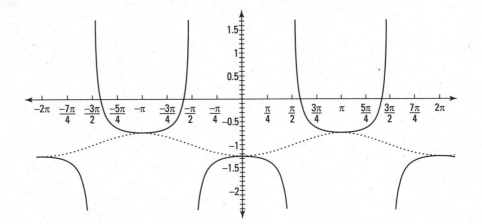

17 Name the period, horizontal shift, and vertical shift of $f(\theta) = 2\sec\left(\dfrac{1}{2}\theta + \dfrac{\pi}{4}\right) + 1$. First, rewrite the function equation: $f(\theta) = 2\sec\dfrac{1}{2}\left(\theta + \dfrac{\pi}{2}\right) + 1$. The period is 4π, the horizontal shift is $-\dfrac{\pi}{2}$, and the vertical shift is 1.

The shifts are evident from the equation. To find the period, simply divide 2π by $\dfrac{1}{2}$ and you get 4π.

18 Graph $f(\theta) = 2\sec\left(\dfrac{1}{2}\theta + \dfrac{\pi}{4}\right) + 1 = 2\sec\dfrac{1}{2}\left(\theta + \dfrac{\pi}{2}\right) + 1$. See the graph for the answer.

Begin by using the information from Question 17 to sketch the reciprocal of the cosine graph. Make appropriate vertical steepening (a factor of 2), a left translation of $\dfrac{\pi}{2}$, and an up-shift of 1.

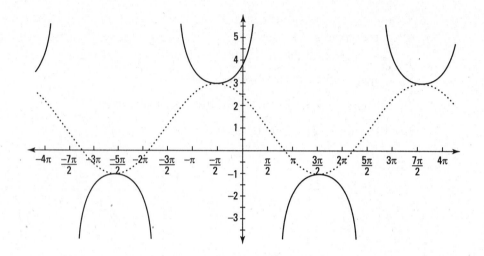

3

Digging into Advanced Trig: Identities, Theorems, and Applications

IN THIS PART . . .

The concepts of trig keep building, and this material is to help you follow along. These chapters move into identities — they're like formulas, but they're true all the time, no matter what you put in for the variable(s). These identities are used to simplify expressions and solve equations, and they're even used in trig proofs (and you thought you were done with proofs in geometry!). This part also includes some very specific and powerful identities that can be used to solve oblique triangles.

Chapter **8**

Basic Trig Identities

Ever want to pretend you were someone else — change your identity? Well, trig expressions have the opportunity to do that all the time. In this chapter, you discover basic identities, or statements that are always true. These identities are used to simplify problems and then to complete trigonometric proofs. Each section builds upon the previous one, so you may want to spend some time reviewing the identities in each section before jumping to the end, to practice proofs.

Using Reciprocal Identities to Simplify Trig Expressions

You actually may have seen some trigonometry expressions back in Chapter 6. In this chapter, you use reciprocal identities to simplify more complicated trig expressions. Because these identities are basically all review, you also find the ratios of tangent and cotangent — the *ratio identities* that are introduced in Chapter 6. The reciprocal (and ratio) identities are:

Reciprocal identities: $\sin x = \dfrac{1}{\csc x}$ or $\csc x = \dfrac{1}{\sin x}$, $\cos x = \dfrac{1}{\sec x}$ or $\sec x = \dfrac{1}{\cos x}$

$\tan x = \dfrac{1}{\cot x}$ or $\cot x = \dfrac{1}{\tan x}$

Ratio identities: $\tan x = \dfrac{\sin x}{\cos x}$ and $\cot x = \dfrac{\cos x}{\sin x}$

Because each pair of expressions is mathematically equivalent, you can substitute one for another in a given expression and watch things simplify. Typically, changing a given expression to all sines and cosines causes a whole lot of canceling! Try it and see!

Q. Use reciprocal identities to rewrite $\dfrac{\sin x \cdot \sec x}{\tan x}$.

 A. The answer is 1. Start by using reciprocal and ratio identities to rewrite $\sec x$ and $\tan x$ in terms of sine and cosine. Next, multiply the two fractions in the numerator. Then, multiply the numerator by the reciprocal of the denominator and reduce before multiplying. Here's what it should look like:

$$\frac{\sin x \cdot \sec x}{\tan x} = \frac{\sin x \cdot \dfrac{1}{\cos x}}{\dfrac{\sin x}{\cos x}} = \frac{\dfrac{\sin x}{\cos x}}{\dfrac{\sin x}{\cos x}} = \frac{\cancel{\sin x}}{\cancel{\cos x}} \cdot \frac{\cancel{\cos x}}{\cancel{\sin x}} = 1$$

---

1 Simplify $\cot x \cdot \sec x$.

2 Simplify $\sin x \cdot \sec x$.

3 Simplify $\sin^3 x \cdot \csc^2 x + \tan x \cdot \cos x$.

4 Simplify $\cot x \cdot \sin x \cdot \tan x$.

Simplifying with Pythagorean Identities

Pythagorean identities are extremely helpful when simplifying complex trig expressions. These identities are derived from those right triangles on a unit circle (turn to Chapter 6 for a review if you need to). Remember that $\cos\theta$ is the x-coordinate on the unit circle and the x leg of a right triangle; $\sin\theta$ is the y-coordinate on the unit circle and the y leg of the right triangle. And the hypotenuse of the triangle on that unit circle is 1. Given the fact that $\text{leg}^2 + \text{leg}^2 = \text{hypotenuse}^2$, you can use that to create the first Pythagorean identity. The other two are derived from that first identity (check out *Pre-Calculus For Dummies* by Mary Jane Sterling [Wiley] if you want to see how this works!). These identities are especially helpful when simplifying expressions that have a term that has been squared ($\sin^2 x$, $\cos^2 x$, and so on). Here are the Pythagorean identities (and some variations):

$$\sin^2 x + \cos^2 x = 1 \quad \text{or} \quad \cos^2 x = 1 - \sin^2 x \quad \text{or} \quad \sin^2 x = 1 - \cos^2 x$$

$$\tan^2 x + 1 = \sec^2 x \quad \text{or} \quad \tan^2 x = \sec^2 - 1 \quad \text{or} \quad 1 = \sec^2 x - \tan^2 x$$

$$\cot^2 x + 1 = \csc^2 x \quad \text{or} \quad \cot^2 x = \csc^2 - 1 \quad \text{or} \quad 1 = \csc^2 x - \cot^2 x$$

Q. Simplify $(\sec x + \tan x)(1 - \sin x)\cos x$.

A. The answer is $\cos^2 x$. Start by changing the terms in the first parentheses using the reciprocal and ratio identities from the preceding section. Then add the resulting fractions (the common denominator is cosine) and cancel the cosine in the numerator and denominator. This leaves you with two terms that you can FOIL. Recognize this last term as a Pythagorean identity? Always handy! Substitute it in and you have your answer. The steps look like this:

$$(\sec x + \tan x)(1 - \sin x)\cos x$$

$$= \left(\frac{1}{\cos x} + \frac{\sin x}{\cos x}\right)(1 - \sin x)\cos x$$

$$= \left(\frac{1 + \sin x}{\cos x}\right)(1 - \sin x)\cancel{\cos x}$$

$$= (1 + \sin x)(1 - \sin x)$$

$$= 1 - \sin^2 x = \cos^2 x$$

---

5 Simplify $\dfrac{\cos x}{\sin x}(\tan x + \cot x)$

6 Simplify $1 - \dfrac{\sin x \cdot \tan x}{\sec x}$

7 Simplify $\sin x \cdot \cot^2 x + \sin x$.

8 Simplify $\left(\sin^2 x - 1\right)\left(\tan^2 x + 1\right)$.

Discovering Even-Odd Identities

All functions, including trig functions, can be described as being even, odd, or neither (see Chapter 3 for review). Knowing whether a trig function is even or odd can actually help you simplify an expression. These even-odd identities are helpful when you have an expression where the variable inside the trig function is negative (such as $-x$). The even-odd identities are:

Even: $\quad \cos(-x) = \cos x \qquad \sec(-x) = \sec x$

Odd: $\quad \begin{aligned} \sin(-x) &= -\sin x \qquad \csc(-x) = -\csc x \\ \tan(-x) &= -\tan x \qquad \cot(-x) = -\cot x \end{aligned}$

Q. Simplify $-\tan^2(-x) + \sec^2(-x)$.

A. The answer is 1. Using the even-odd identities, start by rewriting the statement showing what the squaring does: $-\tan^2(-x) + \sec^2(-x) = -[\tan(-x)]^2 + [\sec(-x)]^2$. Next, substituting for the negative angles: $-[\tan(-x)]^2 + [\sec(-x)]^2 = -[-\tan x]^2 + [\sec x]^2$. Squaring the terms and going back to the original notation, $-[-\tan x]^2 + [\sec x]^2 = -\tan^2 x + \sec^2 x$. Rewrite the expression: $\sec^2 x - \tan^2 x$. Recognize this from our Pythagorean identities from the preceding section? This expression equals 1.

9 Simplify $\sec(-x)\cdot\cot(-x)$.

10 Simplify $\sin x\left[\csc x+\sin(-x)\right]$.

 11 Simplify $\dfrac{\csc x}{\cot(-x)}$.

 12 Simplify $\dfrac{-\cot^2(-x)-1}{-\cot^2 x}$

Simplifying with Co-Function Identities

Ever notice that the graphs of sine and cosine look exactly alike, only shifted? (See Chapter 7 for a visual.) There are also similarities between tangent and cotangent (there is a reflection in addition to a shift), as well as between secant and cosecant. Because these functions have the same values, only shifted, I can define them as being *co-functions*. I can write them as co-function identities and use them to simplify expressions. The co-function identities are:

$$\sin x = \cos\left(\frac{\pi}{2} - x\right) \qquad \cos x = \sin\left(\frac{\pi}{2} - x\right)$$

$$\tan x = \cot\left(\frac{\pi}{2} - x\right) \qquad \cot x = \tan\left(\frac{\pi}{2} - x\right)$$

$$\csc x = \sec\left(\frac{\pi}{2} - x\right) \qquad \sec x = \cos\left(\frac{\pi}{2} - x\right)$$

Q. Simplify $\dfrac{\cot\left(\dfrac{\pi}{2} - x\right)}{\sec x}$

EXAMPLE

A. The answer is $\sin x$. Start by using the co-function identity to replace $\cot\left(\dfrac{\pi}{2} - x\right)$ with $\tan x$. Next, rewrite in terms of sine and cosine using reciprocal and ratio identities. Then, multiply the numerator by the reciprocal of the denominator, canceling any common terms. The steps look like this:

$$\frac{\cot\left(\dfrac{\pi}{2} - x\right)}{\sec x} = \frac{\tan x}{\sec x} = \frac{\dfrac{\sin x}{\cos x}}{\dfrac{1}{\cos x}} = \frac{\sin x}{\cancel{\cos x}} \cdot \frac{\cancel{\cos x}}{1} = \sin x$$

---

13 Simplify $\sin\left(\dfrac{\pi}{2} - x\right) \cdot \cot\left(\dfrac{\pi}{2} - x\right)$

14 Simplify $\sin\left(\dfrac{\pi}{2} - x\right) + \cot\left(\dfrac{\pi}{2} - x\right) \cdot \cos\left(\dfrac{\pi}{2} - x\right)$

15 Simplify $\dfrac{\cos\left(\dfrac{\pi}{2}-x\right)}{\sin\left(\dfrac{\pi}{2}-x\right)}\cdot\cot\left(\dfrac{\pi}{2}-x\right)$

16 Simplify

$$\left[\sec\left(\frac{\pi}{2}-x\right)+\tan\left(\frac{\pi}{2}-x\right)\right]\left[1-\sin\left(\frac{\pi}{2}-x\right)\right]$$

Moving with Periodicity Identities

Recall that horizontal transformations change the period of a graph and horizontal shifts move it left or right (see Chapter 7). If you shift the graph by one whole period to the left or right, you end up with the same function. This is the idea behind *periodicity identities.* Because the periods of sine, cosine, cosecant, and secant repeat every 2π, and tangent and cotangent repeat every π, the periodicity identities are as follows:

$$\sin(x+2\pi)=\sin x \qquad \cos(x+2\pi)=\cos x$$
$$\tan(x+\pi)=\tan x \qquad \cot(x+\pi)=\cot x$$
$$\csc(x+2\pi)=\csc x \qquad \sec(x+2\pi)=\sec x$$

Q. Simplify $1-\sin(2\pi+x)\cdot\cot(x+\pi)\cdot\cos(x+2\pi)$.

EXAMPLE **A.** The answer is $\sin^2 x$. Begin by rewriting the trig terms using periodicity identities: $1-\sin(2\pi+x)\cdot\cot(x+\pi)\cdot\cos(x+2\pi)=1-\sin x\cdot\cot x\cdot\cos x$. Next, rewrite cotangent in terms of sine and cosine: $=1-\sin x\cdot\dfrac{\cos x}{\sin x}\cdot\cos x$. Then, cancel the sine from the numerator and denominator, leaving you with $=1-\sin\!\!\!\diagup\!x\cdot\dfrac{\cos x}{\sin\!\!\!\diagup\!x}\cdot\cos x=1-\cos^2 x$. Using Pythagorean identities, this is the same as $\sin^2 x$.

17 Simplify $\cos(x+2\pi)+\sin(x+2\pi)\cdot\cot(x+\pi)$

18 Simplify $\dfrac{\cos(x+4\pi)}{\cot(x+2\pi)}$

19 Simplify $\dfrac{\sec(x+2\pi)}{\csc(x+2\pi)}$

20 Simplify $\left[\sec(x-2\pi)-\tan(x-\pi)\right]\cdot$ $\left[\sec(x+2\pi)+\tan(x+\pi)\right]$

Tackling Trig Proofs (Identities)

Proofs?!? You thought you left those behind in geometry. Nope, sorry. Don't worry, though — you'll quickly find your way through what is offered here: trig proofs. One thing to remember is that you'll be using what you've already practiced in this chapter. These proofs are composed of two sides of an equation. Your job is to make one side look like the other — that's what makes them an *identity*. Here are some hints on how to solve these:

TIP

≫ **Deal with fractions using basic fraction rules.** The same rules apply to simplifying trig expressions as any other expression. Two key rules to remember:

- Dividing an expression by another expression is the same as multiplying the numerator by the reciprocal of the denominator.

- Use the lowest common denominator (or LCD) when adding or subtracting fractions.

≫ **Factor when you can.** Keep an eye out for factorable terms, including factoring out the greatest common factor (GCF) and factoring trinomials (see Chapter 4).

≫ **Square the square roots.** When you have a square root in a proof, you probably have to square both sides of the equation.

≫ **Work on the more complicated side first.** Because the goal is to make one side look like the other, starting on the more complicated side first is generally the best way to go. The more complicated side usually has more terms in it. If you get stuck, try working on the other side for a while. You can then work backward to simplify the first side.

Q. Prove $\dfrac{1}{\cot^2 x} - \dfrac{1}{\csc^2 x} = \dfrac{\sin^4 x}{\cos^2 x}$

EXAMPLE

A. The left side has more terms, so work on that side. Start by finding the common denominator and adding the fractions.

$$\frac{1}{\cot^2 x} - \frac{1}{\csc^2 x} = \frac{\sin^4 x}{\cos^2 x}$$

$$\frac{\csc^2 x}{\cot^2 x \csc^2 x} - \frac{\cot^2 x}{\cot^2 x \csc^2 x} = \frac{\sin^4 x}{\cos^2 x}$$

$$\frac{\csc^2 x - \cot^2 x}{\cot^2 x \csc^2 x} = \frac{\sin^4 x}{\cos^2 x}$$

From there, notice that you have a Pythagorean identity in the numerator. Substitute in the identity and then the ratio identities; the rewrite the tangent in terms of sine and cosine, and multiply terms to complete the proof.

$$\frac{1}{\cot^2 x \csc^2 x} = \frac{\sin^4 x}{\cos^2 x}$$

$$\tan^2 x \sin^2 x = \frac{\sin^4 x}{\cos^2 x}$$

$$\frac{\sin^2 x}{\cos^2 x} \cdot \sin^2 x = \frac{\sin^4 x}{\cos^2 x}$$

$$\frac{\sin^4 x}{\cos^2 x} = \frac{\sin^4 x}{\cos^2 x}$$

21 Prove $\dfrac{\cot x - 1}{1 + \tan(-x)} = \cot x$ for x in $\left(\dfrac{\pi}{4}, \dfrac{5\pi}{4} \right)$

22 Prove $\dfrac{\csc x + \tan x}{\cot\left(\dfrac{\pi}{2} - x\right)\csc\left(\dfrac{\pi}{2} - x\right)} = \cot^2 x + \cos x$

23 Prove $\cot x = \dfrac{\csc^2 x - 1}{\cot x}$

24 Prove $\dfrac{1 - \sin x}{\csc x} - \dfrac{\tan x \cos x}{1 + \sin x} = -\dfrac{\sin^2 x}{\csc x + 1}$

25 Prove $\sqrt{\left[\cos x \cdot \sin\left(\dfrac{\pi}{2} - x\right) \cdot \csc(2\pi + x) \cdot \sec\left(\dfrac{\pi}{2} - x\right) \right] + 1} = |\csc x|$

26 Prove $\sec x - \cos x = \sin x \cdot \tan x$.

Answers to Problems on Basic Trig Identities

This section contains the answers for the practice problems presented in this chapter. There are also explanations for the answers.

(1) Simplify $\cot x \cdot \sec x$. The answer is $\csc x$.

Start by using ratio and reciprocal identities to rewrite the expression in terms of sine and cosine. Reduce and write as a single fraction. Finally, rewrite using reciprocal identities.

$$\cot x \cdot \sec x = \frac{\cancel{\cos x}}{\sin x} \cdot \frac{1}{\cancel{\cos x}} = \frac{1}{\sin x} = \csc x$$

(2) Simplify $\sin x \cdot \sec x$. The answer is $\tan x$.

Begin by rewriting sec x using a reciprocal identity. Then multiply the terms to get a single fraction. Finally, rewrite as a single expression using a ratio identity.

$$\sin x \cdot \sec x = \sin x \cdot \frac{1}{\cos x} = \frac{\sin x}{\cos x} = \tan x$$

(3) Simplify $\sin^3 x \cdot \csc^2 x + \tan x \cdot \cos x$. The answer is $2\sin x$.

Start by using reciprocal and ratio identities to rewrite the expression in terms of sine and cosine. Next, cancel any terms you can. Finally, combine the like terms.

$$\sin^3 x \cdot \csc^2 x + \tan x \cdot \cos x = \sin^{\cancel{3}1} x \cdot \frac{1}{\cancel{\sin^2 x}} + \frac{\sin x}{\cancel{\cos x}} \cdot \cancel{\cos x} = \frac{\sin x}{1} + \frac{\sin x}{1} = 2\sin x$$

WARNING

Be careful to write $\sin x + \sin x = 2\sin x$, not $\sin 2x$; the latter expression is the sin of a double angle (which is introduced in the next chapter).

(4) Simplify $\cot x \cdot \sin x \cdot \tan x$. The answer is $\sin x$.

Here, you want to start by rewriting the tangent term using ratio identities. Then, cancel the like terms in the numerator and denominator, giving you your answer.

$$\cot x \cdot \sin x \cdot \tan x = \frac{\cancel{\cos x}}{\cancel{\sin x}} \cdot \cancel{\sin x} \cdot \frac{\sin x}{\cancel{\cos x}} = \sin x$$

(5) Simplify $\frac{\cos x}{\sin x}(\tan x + \cot x)$. The answer is $\csc^2 x$.

Begin by using ratio identities to rewrite the tangent and cotangent. Then distribute the fraction, canceling terms in the resulting terms.

$$\frac{\cos x}{\sin x}(\tan x + \cot x) = \frac{\cos x}{\sin x}\left(\frac{\sin x}{\cos x} + \frac{\cos x}{\sin x}\right) = \frac{\cancel{\cos x}}{\sin x} \cdot \frac{\cancel{\sin x}}{\cancel{\cos x}} + \frac{\cos x}{\sin x} \cdot \frac{\cos x}{\sin x} = 1 + \frac{\cos^2 x}{\sin^2 x}$$

Use the ratio identity to change the fraction. This produces a Pythagorean identity.

$$1 + \frac{\cos^2 x}{\sin^2 x} = 1 + \cot^2 x = \csc^2 x$$

6 Simplify $1 - \dfrac{\sin x \cdot \tan x}{\sec x}$. The answer is $\cos^2 x$.

Again, begin by using reciprocal identities to rewrite tangent and secant in terms of sine and cosine. Multiply the factors in the numerator, and then multiply the result by the reciprocal of the denominator. Cancel terms in the numerator and denominator. Finally, use a Pythagorean identity to rewrite the expression as a single term.

$$1 - \frac{\sin x \cdot \tan x}{\sec x} = 1 - \frac{\sin x \cdot \dfrac{\sin x}{\cos x}}{\dfrac{1}{\cos x}} = 1 - \frac{\dfrac{\sin^2 x}{\cos x}}{\dfrac{1}{\cos x}} = 1 - \frac{\sin^2 x}{\cos x} \cdot \frac{\cos x}{1} = 1 - \sin^2 x = \cos^2 x$$

7 Simplify $\sin x \cdot \cot^2 x + \sin x$. The answer is $\csc x$.

Factor $\sin x$ from both terms. Then, using Pythagorean identities, simplify the expression. Next, rewrite the cosecant term using reciprocal identities. Cancel the like terms in the numerator and denominator. Finally, use reciprocal identities to simplify the final term.

$$\sin x \cdot \cot^2 x + \sin x = \sin x \left(\cot^2 x + 1 \right) = \sin x \left(\csc^2 x \right) = \sin x \cdot \frac{1}{\sin^2 x} = \frac{1}{\sin x} = \csc x$$

8 Simplify $\left(\sin^2 x - 1 \right)\left(\tan^2 x + 1 \right)$. The answer is -1.

Start by replacing the first factor using a Pythagorean identity. Note that the replacement will be negative. Next, replace the second factor with a Pythagorean identity. Use a reciprocal identity, reduce the product, and multiply.

$$\left(\sin^2 x - 1 \right)\left(\tan^2 x + 1 \right) = \left(-\cos^2 x \right)\left(\sec^2 x \right) = -\cos^2 x \cdot \frac{1}{\cos^2 x} = -1$$

9 Simplify $\sec(-x) \cdot \cot(-x)$. The answer is $-\csc x$.

Begin by using even-odd identities to get rid of all the $-x$ values inside the trig expressions. Next, use reciprocal and ratio identities to rewrite the expression in terms of sine and cosine. Then, cancel any like terms in the numerator and denominator. Last, rewrite the resulting fraction using reciprocal identities.

$$\sec(-x) \cdot \cot(-x) = \sec x \left(-\cot x \right) = \frac{1}{\cos x} \left(-\frac{\cos x}{\sin x} \right) = -\frac{1}{\sin x} = -\csc x$$

10 Simplify $\sin x \left[\csc x + \sin(-x) \right]$. The answer is $\cos^2 x$.

Use an even-odd identity to replace the $-x$ value; then you can distribute the $\sin x$ term. Using reciprocal identities, rewrite the cosecant term and cancel where you can. The resulting expression can be simplified using a Pythagorean identity.

$$\sin x \left[\csc x + \sin(-x) \right] = \sin x \left[\csc x - \sin x \right] = \sin x \csc x - \sin^2 x = \sin x \cdot \frac{1}{\sin x} - \sin^2 x =$$

$$1 - \sin^2 x = \cos^2 x$$

(11) Simplify $\dfrac{\csc x}{\cot(-x)}$. The answer is $-\sec x$.

Start by replacing the $-x$ value using an even-odd identity. Then, using reciprocal and ratio identities, rewrite everything in terms of sine and cosine. Then multiply the numerator by the reciprocal of the denominator. Cancel any terms you can. Finally, simplify the resulting expression using a reciprocal identity.

$$\frac{\csc x}{\cot(-x)} = \frac{\csc x}{-\cot x} = -\frac{\frac{1}{\sin x}}{\frac{\cos x}{\sin x}} = -\frac{1}{\sin x} \cdot \frac{\sin x}{\cos x} = -\frac{1}{\cos x} = -\sec x$$

(12) Simplify $\dfrac{-\cot^2(-x)-1}{-\cot^2 x}$. The answer is π.

Start by replacing the $-x$ using an even-odd identity. This is a little tricky, because you apply the identity before squaring the replacement. Then simplify. The numerator results in a Pythagorean identity. Next, using reciprocal and ratio identities, change everything into terms using sine and cosine. Multiply the numerator by the reciprocal of the denominator. Cancel what you can and use a reciprocal identity to finish.

$$\frac{-\cot^2(-x)-1}{-\cot^2 x} = \frac{-(\cot(-x))^2 - 1}{-\cot^2 x} = \frac{-(-\cot x)^2 + 1}{-\cot^2 x} = \frac{-\cot^2 x - 1}{-\cot^2 x} = \frac{\cot^2 x + 1}{\cot^2 x} = \frac{\csc^2 x}{\cot^2 x}$$

$$= \frac{\frac{1}{\sin^2 x}}{\frac{\cos^2 x}{\sin^2 x}} = \frac{1}{\sin^2 x} \cdot \frac{\sin^2 x}{\cos^2 x} = \frac{1}{\cos^2 x} = \sec^2 x$$

(13) Simplify $\sin\left(\dfrac{\pi}{2} - x\right) \cdot \cot\left(\dfrac{\pi}{2} - x\right)$. The answer is $\sin x$.

Here you can start by using co-function identities to rewrite the sine and cotangent terms. Next, use a ratio identity to rewrite the tangent term as a rational function of sine and cosine. Cancel the cosine terms, leaving you with your answer.

$$\sin\left(\frac{\pi}{2} - x\right) \cdot \cot\left(\frac{\pi}{2} - x\right) = \cos x \cdot \tan x = \cos x \cdot \frac{\sin x}{\cos x} = \sin x$$

(14) Simplify $\sin\left(\dfrac{\pi}{2} - x\right) + \cot\left(\dfrac{\pi}{2} - x\right) \cdot \cos\left(\dfrac{\pi}{2} - x\right)$. The answer is $\sec x$.

Begin by using co-function identities. Then rewrite tangent using a ratio identity. Next, you need to find a common denominator in order to add the resulting terms. Rewriting the resulting fraction, you can see that the numerator can be simplified using a Pythagorean identity. Finally, use a reciprocal identity to simplify the resulting expression.

$$\sin\left(\frac{\pi}{2} - x\right) + \cot\left(\frac{\pi}{2} - x\right) \cdot \cos\left(\frac{\pi}{2} - x\right) = \cos x + \tan x \cdot \sin x = \cos x + \frac{\sin x}{\cos x} \cdot \sin x = \cos x + \frac{\sin^2 x}{\cos x}$$

$$= \frac{\cos^2 x}{\cos x} + \frac{\sin^2 x}{\cos x} = \frac{\cos^2 x + \sin^2 x}{\cos x} = \frac{1}{\cos x} = \sec x$$

(15) Simplify $\dfrac{\cos\left(\dfrac{\pi}{2}-x\right)}{\sin\left(\dfrac{\pi}{2}-x\right)}\cdot\cot\left(\dfrac{\pi}{2}-x\right)$. The answer is $\tan^2 x$.

Begin by using co-function identities to replace each term. Then use a ratio identity to replace the sine/cosine fraction with tangent. Finally, multiply the tangent terms.

$$\frac{\cos\left(\dfrac{\pi}{2}-x\right)}{\sin\left(\dfrac{\pi}{2}-x\right)}\cdot\cot\left(\frac{\pi}{2}-x\right)=\frac{\sin x}{\cos x}\cdot\tan x=\tan x\cdot\tan x=\tan^2 x$$

(16) Simplify $\left[\sec\left(\dfrac{\pi}{2}-x\right)+\tan\left(\dfrac{\pi}{2}-x\right)\right]\left[1-\sin\left(\dfrac{\pi}{2}-x\right)\right]$. The answer is $\sin x$.

Start by replacing terms using co-function identities. Next, use reciprocal and ratio identities to rewrite the terms using sine and cosine. Then, add the fractions in the first bracket. Next, you can multiply the numerators using FOIL. The resulting numerator can be simplified using a Pythagorean identity. Finally, reduce the fraction.

$$\left[\sec\left(\frac{\pi}{2}-x\right)+\tan\left(\frac{\pi}{2}-x\right)\right]\left[1-\sin\left(\frac{\pi}{2}-x\right)\right]=\left[\csc x+\cot x\right]\left[1-\cos x\right]=\left[\frac{1}{\sin x}+\frac{\cos x}{\sin x}\right]\left[1-\cos x\right]$$

$$=\left[\frac{1+\cos x}{\sin x}\right]\left[\frac{1-\cos x}{1}\right]=\frac{1-\cos^2 x}{\sin x}=\frac{\sin^2 x}{\sin x}=\sin x$$

(17) Simplify $\cos(x+2\pi)+\sin(x+2\pi)\cdot\cot(x+\pi)$. The answer is $2\cos x$.

Using periodicity identities, replace each term. Then, using a ratio identity, replace the cotangent with a ratio of cosine and sine. Reduce and add the two cosines.

$$\cos(x+2\pi)+\sin(x+2\pi)\cdot\cot(x+\pi)=\cos x+\sin x\cdot\cot x=\cos x+\sin x\cdot\frac{\cos x}{\sin x}=$$

$$\cos x+\cos x=2\cos x$$

(18) Simplify $\dfrac{\cos(x+4\pi)}{\cot(x+2\pi)}$. The answer is $\sin x$.

You see twice the fun! In the numerator 4π is a multiple of 2π, and in the denominator 2π is a multiple of π so you really have two periodicity identities. Replacing the terms with the appropriate periodicity identity, you can easily simplify this problem. The next step is to use a ratio identity and then simplify the complex fraction by multiplying the numerator by the reciprocal of the denominator. Cancel to get a simplified expression.

$$\frac{\cos(x+4\pi)}{\cot(x+2\pi)}=\frac{\cos(x+2(2\pi))}{\cot(x+2(\pi))}=\frac{\cos x}{\cot x}=\frac{\cos x}{\dfrac{\cos x}{\sin x}}=\cos x\cdot\frac{\sin x}{\cos x}=\sin x$$

19 Simplify $\dfrac{\sec(x+2\pi)}{\csc(x+2\pi)}$. The answer is $\tan x$.

Start by replacing both terms using periodicity identities. Then, you can rewrite the fraction as a division problem. Next, rewrite the division problem by multiplying by the reciprocal. Then use reciprocal identities to replace both terms. Multiply to write as a single term, which can be simplified using ratio identities.

$$\frac{\sec(x+2\pi)}{\csc(x+2\pi)} = \frac{\sec x}{\csc x} = \frac{\dfrac{1}{\cos x}}{\dfrac{1}{\sin x}} = \frac{1}{\cos x}\cdot\frac{\sin x}{1} = \frac{\sin x}{\cos x} = \tan x$$

20 Simplify $\big[\sec(x-2\pi)-\tan(x-\pi)\big]\cdot\big[\sec(x+2\pi)+\tan(x+\pi)\big]$. The answer is 1.

Simplify by replacing every term using periodicity identities. Then you can FOIL. Finally, use a Pythagorean identity.

$$\big[\sec(x-2\pi)-\tan(x-\pi)\big]\cdot\big[\sec(x+2\pi)+\tan(x+\pi)\big] = [\sec x - \tan x][\sec x + \tan x] =$$
$$\sec^2 x - \tan^2 x = 1$$

21 Prove $\dfrac{\cot x - 1}{1+\tan(-x)} = \cot x$ for x in $\left(\dfrac{\pi}{4}, \dfrac{5\pi}{4}\right)$.

For this proof, you start with the left side because it's more complicated. Begin by using even-odd identities to replace the $-x$ and use ratio identities to rewrite tangent and cotangent in terms of sine and cosine.

$$\frac{\cot x - 1}{1+\tan(-x)} = \frac{\cot x - 1}{1-\tan x} = \frac{\dfrac{\cos x}{\sin x} - 1}{1 - \dfrac{\sin x}{\cos x}} = \cot x$$

Now multiply the numerator and denominator by the LCD of both fractions to simplify the complex fraction.

$$\frac{\dfrac{\cos x}{\sin x} - 1}{1 - \dfrac{\sin x}{\cos x}} = \frac{\dfrac{\cos x}{\sin x} - 1}{1 - \dfrac{\sin x}{\cos x}}\cdot\frac{\sin x \cos x}{\sin x \cos x} = \frac{\dfrac{\cos x}{\sin x}\cdot\sin x \cos x - 1\cdot\sin x \cos x}{1\cdot\sin x\cos x - \dfrac{\sin x}{\cos x}\cdot\sin x \cos x} = \frac{\cos^2 x - \sin x\cos x}{\sin x\cos x - \sin^2 x} = \cot x$$

Finally, you can pull out a common factor on both the numerator and denominator, cancel any like terms, and simplify the resulting fraction using a ratio identity.

$$\frac{\cos^2 x - \sin x\cos x}{\sin x\cos x - \sin^2 x} = \frac{\cos x(\cos x - \sin x)}{\sin x(\cos x - \sin x)} = \frac{\cos x\,\cancel{(\cos x - \sin x)}}{\sin x\,\cancel{(\cos x - \sin x)}} = \frac{\cos x}{\sin x} = \cot x$$

22 Prove $\dfrac{\csc x + \tan x}{\cot\left(\dfrac{\pi}{2}-x\right)\csc\left(\dfrac{\pi}{2}-x\right)} = \cot^2 x + \cos x.$

Start by using co-function identities to replace the terms with π.

$$\frac{\csc x + \tan x}{\cot\left(\dfrac{\pi}{2}-x\right)\csc\left(\dfrac{\pi}{2}-x\right)} = \frac{\csc x + \tan x}{\tan x \sec x} = \cot^2 x + \cos x$$

Next, you can separate the fraction into two different fractions and then cancel any terms on both the numerator and denominator.

$$\frac{\csc x + \tan x}{\tan x \sec x} = \frac{\csc x}{\tan x \sec x} + \frac{\cancel{\tan x}}{\cancel{\tan x} \sec x} = \frac{\csc x}{\tan x \sec x} + \frac{1}{\sec x} = \cot^2 x + \cos x$$

Then, use reciprocal and ratio identities to rewrite all terms using sine and cosine. You can simplify the denominator of the complex fraction using multiplication.

$$\frac{\dfrac{1}{\sin x}}{\dfrac{\sin x}{\cos x} \cdot \dfrac{1}{\cos x}} + \frac{1}{\dfrac{1}{\cos x}} = \frac{\dfrac{1}{\sin x}}{\dfrac{\sin x}{\cos^2 x}} + \frac{1}{\dfrac{1}{\cos x}} = \cot^2 x + \cos x$$

Next, get rid of the complex fractions by multiplying the numerators by the reciprocals of the denominators. The resulting fraction can also be simplified using ratio identities, giving you the answer.

$$\frac{\dfrac{1}{\sin x}}{\dfrac{\sin x}{\cos^2 x}} + \frac{1}{\dfrac{1}{\cos x}} = \frac{1}{\sin x} \cdot \frac{\cos^2 x}{\sin x} + 1 \cdot \frac{\cos x}{1} = \frac{\cos^2 x}{\sin^2 x} + \cos x = \cot^2 x + \cos x$$

(23) Prove $\cot x = \dfrac{\csc^2 x - 1}{\cot x}$

In this proof, the right side is more complicated, so it's wise to start there. Notice the squared term? You can replace it using a Pythagorean identity, which then cancels out the 1 in the numerator. Last, simply cancel a cotangent from both the numerator and denominator, and you're there!

$$\cot x = \frac{\csc^2 x - 1}{\cot x} = \frac{\cot^2 x}{\cot x} = \frac{\cancel{\cot^{2^1}} x}{\cancel{\cot x}} = \cot x$$

(24) Prove $\dfrac{1 - \sin x}{\csc x} - \dfrac{\tan x \cos x}{1 + \sin x} = -\dfrac{\sin^2 x}{\csc x + 1}$

For this proof, you want to start by finding a common denominator for the two fractions and multiplying it through using FOIL for the first fraction. Then you can simplify using a Pythagorean identity.

$$\frac{1 - \sin x}{\csc x} - \frac{\tan x \cos x}{1 + \sin x} = \frac{1 - \sin x}{\csc x} \cdot \frac{1 + \sin x}{1 + \sin x} - \frac{\tan x \cos x}{1 + \sin x} \cdot \frac{\csc x}{\csc x}$$

$$= \frac{1 - \sin^2 x}{\csc x (1 + \sin x)} - \frac{\tan x \cos x \csc x}{\csc x (1 + \sin x)} = \frac{\cos^2 x}{\csc x (1 + \sin x)} - \frac{\tan x \cos x \csc x}{\csc x (1 + \sin x)}$$

$$= -\frac{\sin^2 x}{\csc x + 1}$$

Subtract the two fractions, and then use reciprocal and ratio identities to change the numerator, allowing you to cancel many of the terms. Again, you have a Pythagorean identity that you can simplify.

$$\frac{\cos^2 x - \tan x \cos x \csc x}{\csc x (1 + \sin x)} = \frac{\dfrac{\cos^2 x}{1} - \dfrac{\cancel{\sin x}}{\cancel{\cos x}} \cdot \dfrac{\cancel{\cos x}}{1} \cdot \dfrac{1}{\cancel{\sin x}}}{\csc x (1 + \sin x)} = \frac{\cos^2 x - 1}{\csc x (1 + \sin x)} = -\frac{\sin^2 x}{\csc x + 1}$$

Now that the numerator looks like your final answer, you can concentrate on the denominator. Distribute the cosecant and use reciprocal identities to simplify.

$$\frac{\cos^2 x - 1}{\csc x(1+\sin x)} = \frac{\cos^2 x - 1}{\csc x\left(1+\dfrac{1}{\csc x}\right)} = \frac{\cos^2 x - 1}{\csc x + \dfrac{\csc x}{\csc x}} = -\frac{\sin^2 x}{\csc x + 1}$$

25. Prove $\sqrt{\left[\cos x \cdot \sin\left(\dfrac{\pi}{2}-x\right) \cdot \csc(2\pi + x) \cdot \sec\left(\dfrac{\pi}{2}-x\right)\right]+1} = |\csc x|$

Start by dealing with the gigantic square root by squaring both sides. (**Note:** This method of proof works here because both sides are positive.)

$$\left(\sqrt{\left[\cos x \cdot \sin\left(\dfrac{\pi}{2}-x\right) \cdot \csc(2\pi + x) \cdot \sec\left(\dfrac{\pi}{2}-x\right)\right]+1}\right)^2 = \left(|\csc x|\right)^2$$

$$\left[\cos x \cdot \sin\left(\dfrac{\pi}{2}-x\right) \cdot \csc(2\pi + x) \cdot \sec\left(\dfrac{\pi}{2}-x\right)\right]+1 = \csc^2 x$$

Next, replace any terms with $\dfrac{\pi}{2}$ using co-function identities, and replace any terms with 2π using periodicity identities. Multiply the resulting trig terms.

$$\left[\cos x \cdot \sin\left(\dfrac{\pi}{2}-x\right) \cdot \csc(2\pi + x) \cdot \sec\left(\dfrac{\pi}{2}-x\right)\right]+1 = [\cos x \cdot \cos x \cdot \csc x \cdot \csc x] + 1$$

$$= \cos^2 x \cdot \csc^2 x + 1 = \csc^2 x$$

Replace the cosecant term using reciprocal identities and write as a single fraction that you can replace using a ratio identity.

$$\cos^2 x \cdot \csc^2 x + 1 = \cos^2 x \cdot \frac{1}{\sin^2 x} + 1 = \frac{\cos^2 x}{\sin^2 x} + 1 = \cot^2 x + 1 = \csc^2 x$$

Finally, you have a Pythagorean identity that can be simplified to get the answer you want:

$$\cot^2 x + 1 = \csc^2 x$$

26. Prove $\sec x - \cos x = \sin x \cdot \tan x$.

Start by replacing secant with its reciprocal identity. You end up with a fraction. Add the fractions using a common denominator of cosine and then simplify the resulting numerator with a Pythagorean identity. Finally, rewrite the square of sine as a product. You can then use a ratio identity.

$$\sec x - \cos x = \frac{1}{\cos x} - \cos x = \frac{1}{\cos x} - \frac{\cos^2 x}{\cos x} = \frac{1-\cos^2 x}{\cos x} = \frac{\sin^2 x}{\cos x}$$

$$= \frac{\sin x}{1} \cdot \frac{\sin x}{\cos x} = \sin x \cdot \tan x$$

Chapter **9**

Advanced Trig Identities

O kay . . . it keeps getting better and better. Now you get to use some even more exciting trigonometry — advanced identities, that is. This chapter builds on the basic identities you practice in Chapter 8.

In this chapter, you are given formulas that are essential for calculating precise values of trig functions at certain angles that you can't get any other way (even your calculator only gives approximate answers). These identities are essential for calculus, so it's time to get friendly with advanced identities.

Simplifying with Sum and Difference Identities

Again, you see where mathematicians were able to take simple concepts (addition and subtraction) and relate them to a more complex one (trigonometric angles). These *sum and difference identities* allow you to write an angle that's not from the special triangles of 45-45-90 or 30-60-90 (see Chapter 6) as the sum or difference of those helpful angle measures. For example, you can rewrite the measure of 105° as the sum of 45° and 60°. The problems presented here (and in precalculus books everywhere) can always be written using the angles you already have exact values for, even though these identities can actually be used for any value.

TECHNICAL STUFF

The sum and difference identities are:

$$\sin(a \pm b) = \sin a \cos b \pm \cos a \sin b$$

$$\cos(a \pm b) = \cos a \cos b \mp \sin a \sin b$$

$$\tan(a \pm b) = \frac{\tan a \pm \tan b}{1 \mp \tan a \tan b}$$

EXAMPLE

Q. Find $\tan \dfrac{7\pi}{12}$ using sum or difference identities.

A. $-2 - \sqrt{3}$. Start by breaking up the fraction into a sum of two values that can be found on the unit circle: $\dfrac{7\pi}{12} = \dfrac{3\pi}{12} + \dfrac{4\pi}{12} = \dfrac{\pi}{4} + \dfrac{\pi}{3}$. Next, plug the angles into the sum identity for tangent:

$$\tan\left(\frac{\pi}{4} + \frac{\pi}{3}\right) = \frac{\tan\dfrac{\pi}{4} + \tan\dfrac{\pi}{3}}{1 - \tan\dfrac{\pi}{4}\tan\dfrac{\pi}{3}}$$

WARNING

Keep a close eye on the order of a plus or minus symbol in an equation. If it's inverted to be a minus or plus symbol, then you perform the opposite operation than the given problem. In this

case, the problem involves addition, so you use addition in the numerator and subtraction in the denominator.

Finally, plug in the known values for the angles from the unit circle or chart in Chapter 6 and simplify. Remember to rationalize the denominator using conjugates (see Chapter 2 for review). The steps are as follows:

$$\frac{\tan\dfrac{\pi}{4} + \tan\dfrac{\pi}{3}}{1 - \tan\dfrac{\pi}{4}\tan\dfrac{\pi}{3}} = \frac{1 + \sqrt{3}}{1 - 1 \cdot \sqrt{3}}$$

$$= \frac{1 + \sqrt{3}}{1 - \sqrt{3}} \cdot \frac{1 + \sqrt{3}}{1 + \sqrt{3}}$$

$$= \frac{1 + 2\sqrt{3} + 3}{1 - 3}$$

$$= \frac{4 + 2\sqrt{3}}{-2} = -2 - \sqrt{3}$$

1 Find $\cos 15°$ using sum or difference identities.

2 Express $\tan(45° - x)$ in terms of $\tan x$.

3 Prove $\dfrac{\sin(x+\pi)}{\cos(x+\pi)} = \tan x$.

 4 Prove $\dfrac{\sin(x+y)}{\sin x \cdot \cos y} = 1 + \cot x \cdot \tan y$.

 5 Find $\csc \dfrac{5\pi}{12}$ using sum or difference identities.

 6 Simplify $\sec(180° + x)$ using sum or difference identities.

Using Double-Angle Identities

Double-angle identities help you find the trig value of twice an angle. These can be used to find an exact value if you know the original angle. They can also be used to prove theorems (see Chapter 8) or solve trig equations. Cosine has three double-angle identities created from the identities of Chapter 8. You have a choice as to which you want to use, depending on the problem.

TECHNICAL STUFF

The double-angle identities are:

$$\sin 2a = 2\sin a \cos a$$
$$\cos 2a = \cos^2 a - \sin^2 a = 2\cos^2 a - 1 = 1 - 2\sin^2 a$$
$$\tan 2a = \frac{2\tan a}{1 - \tan^2 a}$$

EXAMPLE

Q. Solve $6\cos^2 x - 6\sin^2 x = 3$ for $0 < x < \pi$.

A. $x = \frac{\pi}{6}$ or $\frac{5\pi}{6}$. Start by factoring out the 6 from the left side of the equation $6(\cos^2 x - \sin^2 x) = 3$ and then divide both sides by 3 to get $2(\cos^2 x - \sin^2 x) = 1$. Next, substitute using the appropriate double-angle identity: $2(\cos 2x) = 1$. Isolate the trigonometric term: $\cos 2x = \frac{1}{2}$. Then, find the inverse of cosine: $2x = \frac{\pi}{3}$ or $2x = \frac{5\pi}{3}$. Finally, solve for x by dividing both sides by 2: $x = \frac{\pi}{6}$ or $\frac{5\pi}{6}$.

---

 7 Find the value of $\cos 2x$ if $\csc x = \frac{12}{5}$.

8 Find the value $\tan 2x$ if $\cot x = \frac{1}{2}$.

9 Prove $\dfrac{\cos 2x + 1}{\sin 2x} = \cot x$.

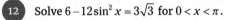 **10** Prove $\dfrac{\cos 2x}{\sin 2x} = \dfrac{1}{2}(\cot x - \tan x)$.

 11 Express $\tan 3x$ in terms of $\tan x$ by using double-angle identities.

12 Solve $6 - 12\sin^2 x = 3\sqrt{3}$ for $0 < x < \pi$.

Reducing with Half-Angle Identities

Similar to sum and difference identities, *half-angle identities* help you find exact values of unusual angles, namely ones that are half the value of ones you already know. For example, if you want to find a trig value of 22.5°, you use the half-angle identity on half of 45° because 22.5 is half of 45. Also, just like every other identity you've reviewed so far, half-angle identities can be used for proving theorems and solving trig equations.

TECHNICAL STUFF

The half-angle identities are:

$$\sin\left(\frac{a}{2}\right) = \pm\sqrt{\frac{1-\cos a}{2}}$$

$$\cos\left(\frac{a}{2}\right) = \pm\sqrt{\frac{1+\cos a}{2}}$$

$$\tan\left(\frac{a}{2}\right) = \frac{1-\cos a}{\sin a} = \frac{\sin a}{1+\cos a}$$

EXAMPLE

Q. Find $\cot\dfrac{5\pi}{12}$ using half-angle identities.

A. $2-\sqrt{3}$. First, because you don't see a half-angle formula for cotangent, you have to start by recognizing that $\cot\dfrac{5\pi}{12}$ is the reciprocal of tangent of the same angle. Therefore, you're going to find the value of $\tan\dfrac{5\pi}{12}$, and then take the reciprocal. The angle $\dfrac{5\pi}{12}$ can be rewritten as $\dfrac{5\pi/6}{2}$. Plugging that into the half-angle identity, you get

$$\tan\left(\frac{5\pi}{12}\right) = \frac{1-\cos\dfrac{5\pi}{6}}{\sin\dfrac{5\pi}{6}}$$

Replacing the trig expressions with the exact values, you get

$$\tan\left(\frac{5\pi}{12}\right) = \frac{1-\left(-\dfrac{\sqrt{3}}{2}\right)}{\dfrac{1}{2}} = \frac{\dfrac{2+\sqrt{3}}{2}}{\dfrac{1}{2}} = 2+\sqrt{3}.$$

But wait — you're not done! This is the value of $\tan\dfrac{5\pi}{12}$, and you need $\cot\dfrac{5\pi}{12}$. So you have to find the reciprocal,

$$\frac{1}{2+\sqrt{3}} = \frac{1}{2+\sqrt{3}}\cdot\frac{2-\sqrt{3}}{2-\sqrt{3}} = \frac{2-\sqrt{3}}{4-3} =$$

$$= \frac{2-\sqrt{3}}{1} = 2-\sqrt{3}.$$

13 Find $\tan\dfrac{3\pi}{8}$ using half-angle identities.

14 Find $\sin\dfrac{7\pi}{12}$.

15 Prove $\left[\tan\left(\dfrac{x}{2}\right)\right]\cdot\tan x = \dfrac{\sin^2 x}{\cos x + \cos^2 x}$.

16 Find an approximate value of $\cos\dfrac{\pi}{24}$ using half-angle identities.

Changing Products to Sums

TECHNICAL STUFF

You need to know three product-to-sum (or product-to-difference) identities: $\sin\cdot\cos$, $\cos\cdot\cos$, and $\sin\cdot\sin$. They follow easily from the sum and difference identities. Here they are:

$$\sin a \cdot \cos b = \tfrac{1}{2}\left[\sin(a+b)+\sin(a-b)\right]$$

$$\cos a \cdot \cos b = \tfrac{1}{2}\left[\cos(a+b)+\cos(a-b)\right]$$

$$\sin a \cdot \sin b = \tfrac{1}{2}\left[\cos(a-b)-\cos(a+b)\right]$$

EXAMPLE

Q. Express $8\sin 3x \cdot \sin x$ as a sum or difference.

A. $4(\cos 2x - \cos 4x)$. Start by plugging into the appropriate product–to–sum identity, sin·sin:

$$8\left(\frac{1}{2}\left[\cos(3x - x) - \cos(3x + x)\right]\right).$$

Then, simplify: $= 4\left[\cos(2x) - \cos(4x)\right]$

 Express $12\cos 6x \cdot \cos 2x$ as a sum or difference.

 Express $2\sin 5x \cdot \cos 2x$ as a sum or difference.

 Express $6\sin 6x \cdot \cos 3x$ as a sum or difference.

 Express $7\sin 8x \cdot \sin 3x$ as a sum or difference.

Expressing Sums as Products

Although less frequently used than the other identities in this chapter, the *sum-to-product identities* are useful for finding exact answers for some trig expressions. In cases where the sum or difference of the two angles results in an angle from the special right triangles (Chapter 6), sum-to-product identities can be quite helpful. The sum-to-product (or difference-to-product) identities involve the addition or subtraction of either sine or cosine. They follow immediately from the product to sum/difference identities by letting $a = \frac{c+d}{2}$ and $b = \frac{c-d}{2}$.

TECHNICAL
STUFF

The sum–to-product (or difference-to-product) identities are

$$\sin c + \sin d = 2\sin\left(\frac{c+d}{2}\right)\cos\left(\frac{c-d}{2}\right) \quad \sin c - \sin d = 2\cos\left(\frac{c+d}{2}\right)\sin\left(\frac{c-d}{2}\right)$$

$$\cos c + \cos d = 2\cos\left(\frac{c+d}{2}\right)\cos\left(\frac{c-d}{2}\right) \quad \cos c - \cos d = -2\sin\left(\frac{c+d}{2}\right)\sin\left(\frac{c-d}{2}\right).$$

Q. Find $\cos 165° + \cos 75°$.

EXAMPLE

A. $-\frac{\sqrt{2}}{2}$. Begin by using the sum-to-product identity $\cos c + \cos d$ to rewrite the expression: $2\cos\left(\frac{165+75}{2}\right)\cos\left(\frac{165-75}{2}\right)$. Simplify the results using unit circle or table values:

$$= 2\cos\left(\frac{240}{2}\right)\cos\left(\frac{90}{2}\right) = 2\cos 120°\cos 45° = 2\left(-\frac{1}{2}\right)\left(\frac{\sqrt{2}}{2}\right) = -\frac{\sqrt{2}}{2}.$$

---

 21 Find $\sin 195° - \sin 75°$.

 **22** Find $\cos 375° - \cos 75°$.

23 Find $\sin\frac{7\pi}{12} + \sin\frac{\pi}{12}$.

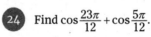 **24** Find $\cos\frac{23\pi}{12} + \cos\frac{5\pi}{12}$.

Powering Down: Power-Reducing Formulas

Power-reducing formulas can be used to simplify trig expressions with exponents and can be used more than once if you have a function that's raised to the fourth power or higher. These nifty formulas help you get rid of exponents; they have many applications in calculus.

TECHNICAL STUFF

Here are the three power-reducing formulas:

$$\sin^2 a = \frac{1-\cos 2a}{2}, \quad \cos^2 a = \frac{1+\cos 2a}{2}, \quad \tan^2 a = \frac{1-\cos 2a}{1+\cos 2a}$$

EXAMPLE

Q. Express $\cos^4 x$ without exponents by using power-reducing formulas.

A. $\frac{1}{8}\left(3+4\cos 2x+\cos 4x\right)$. After rewriting $\cos^4 x$ as $\left(\cos^2 x\right)^2$, you can see that you need to use the power-reducing formula twice. Using it the first time on the function inside the parentheses:

$$\left(\cos^2 x\right)^2 = \left(\frac{1+\cos 2x}{2}\right)^2 = \frac{1+2\cos 2x+\cos^2 2x}{4} = \frac{1}{4}+\frac{\cos 2x}{2}+\frac{\cos^2 2x}{4}$$

Use the formula again on the remaining squared term and reduce:

$$\frac{1}{4}+\frac{\cos 2x}{2}+\frac{\cos^2 2x}{4} = \frac{1}{4}+\frac{\cos 2x}{2}+\frac{1}{4}\left(\frac{1+\cos 2(2x)}{2}\right) = \frac{1}{4}+\frac{\cos 2x}{2}+\frac{1+\cos 4x}{8}$$

Factoring out $\frac{1}{8}$ from each term: $\frac{1}{4}+\frac{\cos 2x}{2}+\frac{1+\cos 4x}{8} = \frac{2}{8}+\frac{4\cos 2x}{8}+\frac{1+\cos 4x}{8} = \frac{1}{8}\left(3+4\cos 2x+\cos 4x\right)$.

---

25 Prove $\cos^2 3x - \sin^2 3x = \cos 6x$.

26 Prove $\left(1+\cos 2x\right)\cdot\tan^3 x = \tan x\cdot\left(1-\cos 2x\right)$.

27 Express $\sin^4 x - \cos^4 x$ without exponents.

28 Express $\cot^2\left(\dfrac{x}{2}\right)$ without exponents.

Answers to Problems on Advanced Trig Identities

Following are the answers to problems dealing with trig functions. You'll also find explanations as to how to find those answers.

(1) Find $\cos 15°$ using sum or difference identities. The answer is $\dfrac{\sqrt{6}+\sqrt{2}}{4}$.

Start by rewriting the angle using special angles from the unit circle or table, letting $15° = 45 - 30°$. Plugging these values into the difference formula, you get: $\cos(45° - \cos 30°) = \cos 45° \cos 30° + \sin 45° \sin 30°$. Use the unit circle or table to plug in the appropriate values and simplify: $= \dfrac{\sqrt{2}}{2} \cdot \dfrac{\sqrt{3}}{2} + \dfrac{\sqrt{2}}{2} \cdot \dfrac{1}{2} = \dfrac{\sqrt{6}}{4} + \dfrac{\sqrt{2}}{4} = \dfrac{\sqrt{6}+\sqrt{2}}{4}$.

(2) Express $\tan(45° - x)$ in terms of $\tan x$. The answer is $\dfrac{1-\tan x}{1+\tan x}$.

Rewrite the expression using the difference formula: $\tan(45° - x) = \dfrac{\tan 45° - \tan x}{1 + \tan 45° \tan x}$. Then, replace $\tan 45°$ with its value of 1 and simplify the expression: $\dfrac{\tan 45° - \tan x}{1 + \tan 45° \tan x} = \dfrac{1-\tan x}{1+1 \cdot \tan x}$

(3) Prove $\dfrac{\sin(x+\pi)}{\cos(x+\pi)} = \tan x$.

Begin the proof by rewriting the left side using sum formulas. Next, substitute in values and simplify. Finally, replace the sine and cosine with tangent using ratio identities (from Chapter 8). The steps are as follows:

$$\dfrac{\sin(x+\pi)}{\cos(x+\pi)} = \dfrac{\sin x \cos \pi + \cos x \sin \pi}{\cos x \cos \pi - \sin x \sin \pi} = \dfrac{\sin x(-1) + \cos x(0)}{\cos x(-1) - \sin x(0)} = \dfrac{-\sin x}{-\cos x} = \tan x$$

(4) Prove $\dfrac{\sin(x+y)}{\sin x \cdot \cos y} = 1 + \cot x \cdot \tan y$.

Start by replacing $\sin(x + y)$ using the sum formula. Next, separate the fraction into the sum of two fractions and reduce terms. Last, rewrite using ratio identities from Chapter 8.

$$\dfrac{\sin(x+y)}{\sin x \cdot \cos y} = \dfrac{\sin x \cos y + \cos x \sin y}{\sin x \cdot \cos y} = \dfrac{\sin x \cos y}{\sin x \cos y} + \dfrac{\cos x \sin y}{\sin x \cos y} = 1 + \dfrac{\cos x}{\sin x} \cdot \dfrac{\sin y}{\cos y} = 1 + \cot x \cdot \tan y$$

(5) Find $\csc \dfrac{5\pi}{12}$ using sum or difference identities. The answer is $\sqrt{6} - \sqrt{2}$.

First, know that you can rewrite $\dfrac{5\pi}{12}$ as $\dfrac{2\pi}{12} + \dfrac{3\pi}{12}$, which is the same as $\dfrac{\pi}{6} + \dfrac{\pi}{4}$. Substitute this back into the problem: $\csc\left(\dfrac{\pi}{6} + \dfrac{\pi}{4}\right)$. Because you don't have a sum identity for cosecant, you need to find $\dfrac{1}{\sin\left(\dfrac{\pi}{6} + \dfrac{\pi}{4}\right)}$. To ease this calculation, you start by finding $\sin\left(\dfrac{\pi}{6} + \dfrac{\pi}{4}\right)$, and then finding the reciprocal of the answer. Using a sum identity, $\sin\left(\dfrac{\pi}{6} + \dfrac{\pi}{4}\right) = \sin\dfrac{\pi}{6}\cos\dfrac{\pi}{4} + \cos\dfrac{\pi}{6}\sin\dfrac{\pi}{4}$. Replace the trig expressions with the appropriate values from the unit circle or table and simplify:

$$=\frac{1}{2}\cdot\frac{\sqrt{2}}{2}+\frac{\sqrt{3}}{2}\cdot\frac{\sqrt{2}}{2}=\frac{\sqrt{2}}{4}+\frac{\sqrt{6}}{4}=\frac{\sqrt{2}+\sqrt{6}}{4}.$$

The last step is to find the reciprocal and simplify by rationalizing the denominator:

$$\csc\left(\frac{\pi}{6}+\frac{\pi}{4}\right)=\frac{4}{\sqrt{2}+\sqrt{6}}\cdot\frac{\sqrt{2}-\sqrt{6}}{\sqrt{2}-\sqrt{6}}=\frac{4\left(\sqrt{2}-\sqrt{6}\right)}{2-6}=\frac{\cancel{4}\left(\sqrt{2}-\sqrt{6}\right)}{-\cancel{4}}=\sqrt{6}-\sqrt{2}$$

⑥ Simplify $\sec(180°+x)$ using sum or difference identities. The answer is $-\sec x$.

Because you don't have a sum identity for secant, you need to use the one for cosine and then find the reciprocal. Using the cosine identity, you get $\cos(180°+x)=\cos180°\cos x-\sin180°\sin x$. Plugging in values and simplifying, you get $=(-1)\cos x-(0)\sin x=-\cos x$. The reciprocal is

$\dfrac{1}{-\cos x}=-\sec x$. See Chapter 8 for more on reciprocal identities.

⑦ Find the value of $\cos2x$ if $\csc x=\dfrac{12}{5}$. The answer is $\dfrac{47}{72}$.

Because $\cos2x=1-2\sin^2 x$, you need to know $\sin x$ to plug it in. No problem! You have the reciprocal: $\csc x$. The value of the reciprocal then is, $\sin x=\dfrac{5}{12}$. Plugging this in to the

identity, you get $\cos2x=1-2\left(\dfrac{5}{12}\right)^2=1-\cancel{2}\left(\dfrac{25}{\underset{72}{\cancel{144}}}\right)=1-\dfrac{25}{72}=\dfrac{47}{72}$.

⑧ Find the value of $\tan2x$ if $\cot x=\dfrac{1}{2}$. The answer is $-\dfrac{4}{3}$.

Begin by finding $\tan x$ by taking the reciprocal of $\cot x$: $\tan x=\dfrac{1}{\frac{1}{2}}=2$. Then plug it into the

formula for $\tan2x$: $\tan2x=\dfrac{2\tan x}{1-\tan^2 x}=\dfrac{2(2)}{1-2^2}=\dfrac{4}{1-4}=-\dfrac{4}{3}$

⑨ Prove $\dfrac{\cos2x+1}{\sin2x}=\cot x$.

Working with the left side, change both double angles using the double-angle identities. After combining like terms, reduce the fraction. Finally, rewrite the result using ratio identities (see Chapter 8). The steps are as follows:

$$\frac{\cos2x+1}{\sin2x}=\frac{2\cos^2 x-1+1}{2\sin x\cos x}=\frac{\cancel{2}\cos^{\cancel{2}^{1}} x}{\cancel{2}\sin x\cancel{\cos x}}=\frac{\cos x}{\sin x}=\cot x$$

⑩ Prove $\dfrac{\cos2x}{\sin2x}=\dfrac{1}{2}(\cot x-\tan x)$.

Begin by using the double-angle identities for the left side: $\dfrac{\cos2x}{\sin2x}=\dfrac{\cos^2 x-\sin^2 x}{2\sin x\cos x}=$
$\dfrac{1}{2}(\cot x-\tan x)$. Next, separate the single fraction into the difference of two fractions:

$\dfrac{\cos^2 x}{2\sin x\cos x}-\dfrac{\sin^2 x}{2\sin x\cos x}=\dfrac{1}{2}(\cot x-\tan x)$. Then, cancel any terms in the numerator and

denominator: $\dfrac{\cos^{\cancel{2}^{1}} x}{2\sin x\cancel{\cos x}}-\dfrac{\sin^{\cancel{2}^{1}} x}{2\cancel{\sin x}\cos x}=\dfrac{\cos x}{2\sin x}-\dfrac{\sin x}{2\cos x}=\dfrac{1}{2}(\cot x-\tan x)$.

Finally, use ratio identities to replace the fractions with cotangent and tangent and factor out the GCF: $\dfrac{\cos x}{2\sin x} - \dfrac{\sin x}{2\cos x} = \dfrac{1}{2}\cot x - \dfrac{1}{2}\tan x = \dfrac{1}{2}(\cot x - \tan x)$.

11 Express $\tan 3x$ in terms of $\tan x$ by using double-angle identities. The answer is $\dfrac{3\tan x - \tan^3 x}{1 - 3\tan^2 x}$.

Start by separating out the angle into a single and double angle: $\tan(x + 2x)$. Next, use the sum identity for tangent: $\tan(x + 2x) = \dfrac{\tan x + \tan 2x}{1 - \tan x \tan 2x}$. Next, replace the double-angle terms

using the double-angle formula: $\dfrac{\tan x + \tan 2x}{1 - \tan x \tan 2x} = \dfrac{\tan x + \dfrac{2\tan x}{1 - \tan^2 x}}{1 - \tan x \cdot \dfrac{2\tan x}{1 - \tan^2 x}}$. Then, to simplify

the complex fraction, multiply by the common denominator of the two fractions:

$$\dfrac{\tan x + \dfrac{2\tan x}{1 - \tan^2 x}}{1 - \tan x \cdot \dfrac{2\tan x}{1 - \tan^2 x}} \cdot \dfrac{1 - \tan^2 x}{1 - \tan^2 x} = \dfrac{\tan x(1 - \tan^2 x) + \dfrac{2\tan x}{1 - \tan^2 x} \cdot (1 - \tan^2 x)}{1(1 - \tan^2 x) - \tan x \cdot \dfrac{2\tan x}{1 - \tan^2 x} \cdot (1 - \tan^2 x)} = \dfrac{\tan x(1 - \tan^2 x) + 2\tan x}{1 - \tan^2 x - 2\tan^2 x}$$

Multiply and combine like terms: $\dfrac{\tan x(1 - \tan^2 x) + 2\tan x}{1 - 3\tan^2 x} = \dfrac{\tan x - \tan^3 x + 2\tan x}{1 - 3\tan^2 x} =$

$\dfrac{3\tan x - \tan^3 x}{1 - 3\tan^2 x}$

12 Solve $6 - 12\sin^2 x = 3\sqrt{3}$ for $0 < x < \pi$. The answer is $\dfrac{\pi}{12}$ or $\dfrac{11\pi}{12}$.

Begin by factoring out 6 from both terms on the left and dividing both sides by 3: $6(1 - 2\sin^2 x) = 3\sqrt{3}$ and then $2(1 - 2\sin^2 x) = \sqrt{3}$. Next, use a double-angle identity to replace the trigonometric term: $2(\cos 2x) = \sqrt{3}$]. Isolate the trig term by dividing both sides by 2: $\cos 2x = \dfrac{\sqrt{3}}{2}$. Then, to solve for x, take the inverse cosine of each side: $\cos^{-1}(\cos 2x) = \cos^{-1}\left(\dfrac{\sqrt{3}}{2}\right)$ becomes

$2x = \cos^{-1}\left(\dfrac{\sqrt{3}}{2}\right)$. And $2x = \dfrac{\pi}{6}, \dfrac{11\pi}{6}$. Dividing by 2, $x = \dfrac{\pi}{12}, \dfrac{11\pi}{12}$

13 Find $\tan \dfrac{3\pi}{8}$ using half-angle identities. The answer is $\sqrt{2} + 1$.

Start by rewriting the angle so you can use the half angle identity: $\dfrac{3\pi}{8} = \dfrac{3\pi/4}{2}$. Then, plug it into the half-angle identity: $\tan\left(\dfrac{3\pi/4}{2}\right) = \dfrac{1 - \cos 3\pi/4}{\sin 3\pi/4}$. Find the exact values for the

trigonometric terms: $\dfrac{1 - \cos 3\pi/4}{\sin 3\pi/4} = \dfrac{1 - \left(-\dfrac{\sqrt{2}}{2}\right)}{\dfrac{\sqrt{2}}{2}}$. Then simplify the complex fraction by multi-

plying the numerator by the reciprocal of the denominator: $\dfrac{1 - \left(-\dfrac{\sqrt{2}}{2}\right)}{\dfrac{\sqrt{2}}{2}} = \dfrac{1 + \dfrac{\sqrt{2}}{2}}{\dfrac{\sqrt{2}}{2}} = \dfrac{\dfrac{2}{2} + \dfrac{\sqrt{2}}{2}}{\dfrac{\sqrt{2}}{2}} =$

$\dfrac{2 + \sqrt{2}}{2} \cdot \dfrac{2}{\sqrt{2}} = \dfrac{2 + \sqrt{2}}{\sqrt{2}}$. Now rationalize and reduce: $\dfrac{2 + \sqrt{2}}{\sqrt{2}} \cdot \dfrac{\sqrt{2}}{\sqrt{2}} = \dfrac{2\sqrt{2} + 2}{2} = \sqrt{2} + 1$

(14) Find $\sin\dfrac{7\pi}{12}$. The answer is $\dfrac{\sqrt{2+\sqrt{3}}}{2}$.

Begin by rewriting the angle: $\dfrac{7\pi}{12} = \dfrac{7\pi/6}{2}$. Plug the angle into the appropriate half-angle

identity: $\sin\left(\dfrac{7\pi/6}{2}\right) = \sqrt{\dfrac{1-\cos 7\pi/6}{2}}$. Because the angle is in Quadrant II, sine will be positive.

The next step is to replace the trig term with the function value: $\sqrt{\dfrac{1-\cos 7\pi/6}{2}} = \sqrt{\dfrac{1-\left(-\dfrac{\sqrt{3}}{2}\right)}{2}}$.

Simplify the complex fraction: $\sqrt{\dfrac{1-\left(-\dfrac{\sqrt{3}}{2}\right)}{2}} = \sqrt{\dfrac{1+\dfrac{\sqrt{3}}{2}}{2}} = \sqrt{\dfrac{\dfrac{2+\sqrt{3}}{2}}{2}} = \sqrt{\dfrac{2+\sqrt{3}}{4}} = \dfrac{\sqrt{2+\sqrt{3}}}{\sqrt{4}} = \dfrac{\sqrt{2+\sqrt{3}}}{2}$.

(15) Prove $\left[\tan\left(\dfrac{x}{2}\right)\right]\cdot\tan x = \dfrac{\sin^2 x}{\cos x + \cos^2 x}$.

Start with the left side by replacing the half-angle term using the appropriate identity:
$\left[\tan\left(\dfrac{x}{2}\right)\right]\cdot\tan x = \dfrac{\sin x}{1+\cos x}\cdot\tan x = \dfrac{\sin^2 x}{\cos x + \cos^2 x}$. Next, use a ratio identity to change
tangent to sine and cosine and multiply through to complete the proof: $\dfrac{\sin x}{1+\cos x}\cdot\tan x = $
$\dfrac{\sin x}{1+\cos x}\cdot\dfrac{\sin x}{\cos x} = \dfrac{\sin^2 x}{\cos x + \cos^2 x}$

(16) Find an approximate value of $\cos\dfrac{\pi}{24}$ using half-angle identities. The answer is

$\dfrac{\sqrt{2+\sqrt{2+\sqrt{3}}}}{2} \approx 0.99$.

Rewriting the angle to use in a half-angle identity, remember that the angle is in

Quadrant I, so the cosine is positive: $\dfrac{\pi}{24} = \dfrac{\pi/12}{2}$. Use the appropriate half-angle identity:

$\cos\left(\dfrac{\pi/12}{2}\right) = \sqrt{\dfrac{1+\cos\dfrac{\pi}{12}}{2}}$. Because you don't have a special right triangle value, you need to

use a half-angle identity again: $\sqrt{\dfrac{1+\cos\dfrac{\pi}{12}}{2}} = \sqrt{\dfrac{1+\cos\dfrac{\pi/6}{2}}{2}} = \sqrt{\dfrac{1+\sqrt{\dfrac{1+\cos\dfrac{\pi}{6}}{2}}}{2}}$. Now you can

replace the cosine term using the function value (finally!): $\sqrt{\dfrac{1+\sqrt{\dfrac{1+\cos\dfrac{\pi}{6}}{2}}}{2}} = \sqrt{\dfrac{1+\sqrt{\dfrac{1+\dfrac{\sqrt{3}}{2}}{2}}}{2}}$.

Finish by simplifying the expression:

$\sqrt{\dfrac{1+\sqrt{\dfrac{1+\dfrac{\sqrt{3}}{2}}{2}}}{2}} = \sqrt{\dfrac{1+\sqrt{\dfrac{\dfrac{2+\sqrt{3}}{2}}{2}}}{2}} = \sqrt{\dfrac{1+\sqrt{\dfrac{2+\sqrt{3}}{4}}}{2}} = \sqrt{\dfrac{1+\dfrac{\sqrt{2+\sqrt{3}}}{2}}{2}} = \sqrt{\dfrac{\dfrac{2+\sqrt{2+\sqrt{3}}}{2}}{2}}$

$= \sqrt{\dfrac{2+\sqrt{2+\sqrt{3}}}{4}} = \dfrac{\sqrt{2+\sqrt{2+\sqrt{3}}}}{2}$

This comes out to be approximately 0.99.

(17) Express $12\cos 6x \cdot \cos 2x$ as a sum or difference. The answer is $6(\cos 8x + \cos 4x)$.

Since you see cosine times cosine, rewrite using the product-to-sum identity $\cos a \cdot \cos b = \frac{1}{2}[\cos(a+b) + \cos(a-b)]$. In this case $12\cos 6x \cdot \cos 2x = 12\left(\frac{1}{2}[\cos(6x+2x) + \cos(6x-2x)]\right) = 6[\cos 8x + \cos 4x]$

(18) Express $2\sin 5x \cdot \cos 2x$ as a sum or difference. The answer is $\sin 7x + \sin 3x$.

Begin by plugging into the appropriate product-to-sum identity: $\sin a \cdot \cos b = \frac{1}{2}[\sin(a+b) + \sin(a-b)]$.

Then simplify: $2\sin 5x \cdot \cos 2x = 2\left(\frac{1}{2}[\sin(5x+2x) + \sin(5x-2x)]\right) = \sin 7x + \sin 3x$.

(19) Express $6\sin 6x \cdot \cos 3x$ as a sum or difference. The answer is $3(\sin 9x + \sin 3x)$.

Again, plug in the appropriate product-to-sum identity: $\sin a \cdot \cos b = \frac{1}{2}[\sin(a+b) + \sin(a-b)]$.

Then simplify: $6\sin 6x \cdot \cos 3x = 6\left(\frac{1}{2}[\sin(6x+3x) + \sin(6x-3x)]\right) = 3[\sin 9x + \sin 3x]$.

(20) Express $7\sin 8x \cdot \sin 3x$ as a sum or difference. The answer is $\frac{7}{2}(\cos 5x - \cos 11x)$.

You guessed it! Plug in the appropriate product-to-sum identity: $\sin a \cdot \sin b = \frac{1}{2}[\cos(a-b) - \cos(a+b)]$.

Simplify: $7\sin 8x \cdot \sin 3x = 7\left(\frac{1}{2}[\cos(8x-3x) - \cos(8x+3x)]\right) = \frac{7}{2}[\cos 5x - \cos 11x]$.

(21) Find $\sin 195° - \sin 75°$. The answer is $-\frac{\sqrt{6}}{2}$.

Begin by using the sum-to-product identity $\sin c - \sin d = 2\cos\left(\frac{c+d}{2}\right)\sin\left(\frac{c-d}{2}\right)$ to rewrite the expression: $\sin 195° - \sin 75° = 2\cos\left(\frac{195+75}{2}\right)\sin\left(\frac{195-75}{2}\right) = 2\cos(135)\sin(120)$. Simplify the result using unit circle values: $2\cos(135°)\sin(120°) = 2\left(-\frac{\sqrt{2}}{2}\right)\left(\frac{\sqrt{3}}{2}\right) = -\frac{\sqrt{6}}{2}$

(22) Find $\cos 375° - \cos 75°$. The answer is $\frac{\sqrt{2}}{2}$.

Use the sum-to-product identity $\cos c - \cos d = -2\sin\left(\frac{c+d}{2}\right)\sin\left(\frac{c-d}{2}\right)$ to rewrite the expression: $\cos 375° - \cos 75° = -2\sin\left(\frac{375+75}{2}\right)\sin\left(\frac{375-75}{2}\right) = -2\sin(225)\sin(150)$. Then, plug in the values and simplify: $= -2\left(-\frac{\sqrt{2}}{2}\right)\left(\frac{1}{2}\right) = \frac{\sqrt{2}}{2}$

(23) Find $\sin\frac{7\pi}{12} + \sin\frac{\pi}{12}$. The answer is $\frac{\sqrt{6}}{2}$.

Rewrite the expression with the appropriate sum-to-product identity:

$\sin\frac{7\pi}{12} + \sin\frac{\pi}{12} = 2\sin\left(\frac{\frac{7\pi}{12} + \frac{\pi}{12}}{2}\right)\cos\left(\frac{\frac{7\pi}{12} - \frac{\pi}{12}}{2}\right) = 2\sin\left(\frac{8\pi}{24}\right)\cos\left(\frac{6\pi}{24}\right) = 2\sin\left(\frac{\pi}{3}\right)\cos\left(\frac{\pi}{4}\right)$

Substitute in the values to simplify the result: $2\sin\left(\frac{\pi}{3}\right)\cos\left(\frac{\pi}{4}\right) = 2\left(\frac{\sqrt{3}}{2}\right)\left(\frac{\sqrt{2}}{2}\right) = \frac{\sqrt{6}}{2}$

(24) Find $\cos\dfrac{23\pi}{12} + \cos\dfrac{5\pi}{12}$. The answer is $\dfrac{\sqrt6}{2}$.

Again, start by using a sum-to-product identity to rewrite the expression:

$$\cos\frac{23\pi}{12} + \cos\frac{5\pi}{12} = 2\cos\left(\frac{\frac{23\pi}{12}+\frac{5\pi}{12}}{2}\right)\cos\left(\frac{\frac{23\pi}{12}-\frac{5\pi}{12}}{2}\right) = 2\cos\left(\frac{28\pi}{24}\right)\cos\left(\frac{18\pi}{24}\right) = 2\cos\left(\frac{7\pi}{6}\right)\cos\left(\frac{3\pi}{4}\right).$$

Next, simplify the result using function values:

$$2\cos\left(\frac{7\pi}{6}\right)\cos\left(\frac{3\pi}{4}\right) = 2\left(-\frac{\sqrt3}{2}\right)\left(-\frac{\sqrt2}{2}\right) = \frac{\sqrt6}{2}$$

(25) **Prove** $\cos^2 3x - \sin^2 3x = \cos 6x$.

Working on the left side, use the power-reducing formulas to rewrite both terms:

$$\cos^2 3x - \sin^2 3x = \frac{1+\cos 2(3x)}{2} - \frac{1-\cos 2(3x)}{2} = \frac{1+\cos 6x}{2} - \frac{1-\cos 6x}{2} = \cos 6x.$$ Combine

the fractions into one and combine like terms: $\dfrac{1+\cos 6x}{2} - \dfrac{1-\cos 6x}{2} = \dfrac{1+\cos 6x - (1-\cos 6x)}{2} =$

$\dfrac{2\cos 6x}{2} = \cos 6x.$

(26) **Prove** $(1+\cos 2x)\cdot\tan^3 x = \tan x\cdot(1-\cos 2x)$.

Start by rewriting the tangent term on the left so that you can use a power-reducing formula: $(1+\cos 2x)\cdot\tan x\cdot\tan^2 x = \tan x\cdot(1-\cos 2x)$. Now use the power-reducing formula:

$(1+\cos 2x)\cdot\tan x\cdot\tan^2 x = (1+\cos 2x)\cdot\tan x\cdot\dfrac{1-\cos 2x}{1+\cos 2x} = \tan x\cdot(1-\cos 2x)$. Cancel terms in

the numerator and denominator: $\dfrac{(1+\cos 2x)}{1}\cdot\tan x\cdot\dfrac{1-\cos 2x}{1+\cos 2x} = \tan x\cdot(1-\cos 2x)$, and you're there!

(27) Express $\sin^4 x - \cos^4 x$ without exponents. The answer is $-\cos 2x$.

For this one, you start by factoring the difference of two squares: $\sin^4 x - \cos^4 x = \left(\sin^2 x + \cos^2 x\right)$ $\left(\sin^2 x - \cos^2 x\right)$. Now use a Pythagorean identity (see Chapter 8) to simplify: $= (1)\left(\sin^2 x - \cos^2 x\right)$. If you replace the cosine term using another version of that Pythagorean identity, you can combine like terms to have only one squared term remaining: $= \left(\sin^2 x - \left(1-\sin^2 x\right)\right) = \left(\sin^2 x - 1 + \sin^2 x\right) = 2\sin^2 x - 1$. From here, you just need to replace the squared term using power-reducing formulas and simplify: $2\sin^2 x - 1 = 2\left(\dfrac{1-\cos 2x}{2}\right) - 1 = 1-\cos 2x - 1 = -\cos 2x$.

(28) Express $\cot^2\left(\dfrac{x}{2}\right)$ without exponents. The answer is $\dfrac{1+\cos x}{1-\cos x}$.

Because you don't have a power-reducing formula for cotangent, you need to start by using ratio identities (see Chapter 8) to rewrite the expression: $\cot^2\left(\dfrac{x}{2}\right) = \dfrac{1}{\tan^2\left(\dfrac{x}{2}\right)}$. Insert the

power-reducing identity: $\dfrac{1}{\tan^2\left(\dfrac{x}{2}\right)} = \dfrac{1}{\dfrac{1-\cos 2\left(\frac{x}{2}\right)}{1+\cos 2\left(\frac{x}{2}\right)}} = \dfrac{1}{\dfrac{1-\cos x}{1+\cos x}}$ and then multiply by the

reciprocal of the denominator: $\dfrac{1}{\dfrac{1-\cos x}{1+\cos x}} \cdot \dfrac{\dfrac{1+\cos x}{1-\cos x}}{\dfrac{1+\cos x}{1-\cos x}} = \dfrac{\dfrac{1+\cos x}{1-\cos x}}{\dfrac{1-\cos^2 x}{1-\cos^2 x}} = \dfrac{1+\cos x}{1-\cos x}.$

Chapter **10**

Solving Oblique Triangles

The trigonometry functions sine, cosine, and tangent are great for finding missing sides and angles inside right triangles. But what happens when a triangle isn't quite right? This type of triangle is known as an oblique triangle — any kind of triangle that isn't a right triangle. As you see in Chapter 6, the process of finding all the sides and angles in a triangle is known as solving the triangle. This chapter helps you figure out that process for oblique triangles.

As long as you know one angle and the side directly across from it (plus one more piece of information), you can use the Law of Sines to solve the triangle. The Law of Sines can be used in three different cases: angle-side-angle (ASA), angle-angle-side (AAS), and side-side-angle (SSA). The first two cases have exactly one solution. The third case is known as *the ambiguous case* because it may have one, two, or no solutions. You'll see an explanation of each case to show you how to deal with them. In fact, the ambiguous case gets its own section in this chapter.

If you have information, but it doesn't include an angle and the side opposite it, you can start off with the Law of Cosines, which can be used for two cases: side-side-side (SSS) and side-angle-side (SAS).

TIP Most books use *standard notation* to label an oblique triangle: Each vertex is labeled with a capital letter, and the side opposite it is the same lowercase letter (across from angle A is side *a*, and so on). It's recommended that you draw the triangle and label it with the information you're given. From the figure, you can tell which case you've got on your hands and, therefore, which formula to use.

Solving a Triangle with the Law of Sines: ASA and AAS

After you draw out your triangle and see that you have two angles and the side in between them (ASA) or two angles and a consecutive side (AAS), proceed to solve the triangle with the Law of Sines:

$$\frac{a}{\sin A} = \frac{b}{\sin B} = \frac{c}{\sin C}$$

TECHNICAL STUFF In either case, because you know two angles inside the given triangle, you can compute the third, because the sum of the angles inside any triangle is always 180°. Remember, you'll find exactly one solution in either of these cases of the Law of Sines.

Q. Solve the triangle ABC if A = 54°, B = 28°, and c = 11.2.

EXAMPLE

A. C = 98°, a ≈ 9.15, b ≈ 5.31. First, draw and label the triangle. You can see that you have a side sandwiched between two angles (ASA), so you know you can start with the Law of Sines. Because you know two of the angles in the triangle, you can find the third angle: $180° - (54° + 28°) = 98°$; C = 98°. Now substitute all the given information and the angle you just found into the Law of Sines: $\frac{a}{\sin 54°} = \frac{b}{\sin 28°} = \frac{11.2}{\sin 98°}$. Using the second two equivalent ratios gives you a proportion you can solve: $\frac{b}{\sin 28°} = \frac{11.2}{\sin 98°}$. Cross multiply to get $b \cdot \sin 98° = 11.2 \cdot \sin 28°$. Divide both sides by sin 98° to get $b = \frac{11.2 \cdot \sin 28°}{\sin 98°}$. Use your calculator to get b ≈ 5.31.

Note: If you put trig function values into your calculator and round as you go, your final answer will be affected. The best course of action is to wait until the end; the answer will be more precise.

Set the first and third ratios equal to each other to get the proportion that can be solved for a: $\frac{a}{\sin 54°} = \frac{11.2}{\sin 98°}$. Cross-multiply to get $a \cdot \sin 98° = 11.2 \cdot \sin 54°$. Divide both sides by sin 98° to get $a = \frac{11.2 \cdot \sin 54°}{\sin 98°}$ or a ≈ 9.15.

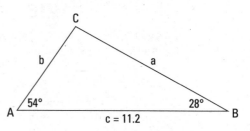

Solve the triangle if B = 46°, C = 62°, and $a = 21$.

 Solve the triangle if A = 19°, C = 100°, and $b = 4.4$.

 Solve the triangle if A = 49°, B = 21°, and $a = 5$.

 Solve the triangle if A = 110°, C = 56°, and $a = 8$.

Tackling Triangles in the Ambiguous Case: SSA

A triangle with two sides and a consecutive angle is the *ambiguous case* of the Law of Sines. In this case, you may have one, two, or no solutions. It's usually best to assume that two solutions exist. That way, in attempting to find them both, you'll discover, when doing the computations, that there are no solutions, or one solution. Or, you'll find one solution and then find a second one that works (so there really are two solutions).

TIP When using your calculator to solve these types of problems, anytime you use an inverse trig function (like $\sin^{-1}$, $\cos^{-1}$, or $\tan^{-1}$) to solve for an angle, know that the calculator may give you the reference angle, or the first quadrant answer θ (for more review on this, check out Chapter 6). There might be a second quadrant answer, $180 - \theta$.

EXAMPLE

Q. Solve the triangle if $a = 25$, $c = 15$, and $C = 40°$.

A. No solution. Start off by substituting the given information into the Law of Sines: $\dfrac{25}{\sin A} = \dfrac{b}{\sin B} = \dfrac{15}{\sin 40°}$. Notice that the middle ratio has absolutely no information in it at all, so you basically ignore it for now and work with the first and third ratios to get the proportion: $\dfrac{25}{\sin A} = \dfrac{15}{\sin 40°}$.

Now, cross-multiply to get $25 \cdot \sin 40° = 15 \cdot \sin A$. Solve for the trig function with the variable in it to get $\dfrac{25 \cdot \sin 40°}{15} = \sin A$. Put the expression into your calculator to get that $\sin A \approx 1.07$. Even if you momentarily forget that sine has only values between -1 and 1 inclusively and you inverse sine both sides of the equation, you'll get an error message on your calculator that tells you there's no solution.

---

5 Solve the triangle if $b = 8$, $c = 14$, and $C = 37°$.

6 Solve the triangle if $b = 5$, $c = 12$, and $B = 20°$.

7 Solve the triangle if $a = 10$, $c = 24$, and $A = 102°$.

8 Solve the triangle if $b = 8$, $c = 16$, and $B = 30°$.

Conquering a Triangle with the Law of Cosines: SAS and SSS

The Law of Cosines comes in handy when the Law of Sines doesn't work. Specifically, you use the Law of Cosines in two cases:

>> You know two sides and the angle in between them (SAS).

>> You know all three sides (SSS).

TECHNICAL STUFF

The Law of Cosines is:

$$a^2 = b^2 + c^2 - 2bc\cos A \qquad b^2 = a^2 + c^2 - 2ac\cos B \qquad c^2 = a^2 + b^2 - 2ab\cos C.$$

You may also have seen the three forms presented in terms of each of the angles in the triangle:

$$A = \cos^{-1}\left(\frac{b^2 + c^2 - a^2}{2bc}\right) \quad B = \cos^{-1}\left(\frac{a^2 + c^2 - b^2}{2ac}\right) \quad C = \cos^{-1}\left(\frac{a^2 + b^2 - c^2}{2ab}\right)$$

TIP

You don't have to memorize all these formulas. The ones for the angles are just the first three, each rewritten in terms of the angle. You can find the steps for developing one of them in *Pre-Calculus For Dummies* by Mary Jane Sterling (Wiley), so if you'd like to see exactly how you arrive at the angle formulas, check it out there.

EXAMPLE

Q. Solve the triangle if $A = 40°$, $b = 10$, and $c = 7$.

A. $a \approx 6.46$, $C \approx 44.1°$, $B \approx 95.9°$. If you draw the triangle, you notice that this time the given angle is between two sides (SAS), and you know that you have to use the Law of Cosines. Substituting in: $a^2 = 10^2 + 7^2 - 2(10)(7)\cos 40°$. Using a calculator you get $a^2 \approx 41.75377796$, or $a \approx 6.46$. (Remember to keep as many decimal places until you get to your final answer.) The value of a that shows on the calculator, before rounding, is 6.461716333. Next, continue using the Law of Cosines to solve angle C and get $7^2 = 41.75377796 + 10^2 - 2(6.461716333)(10)\cos C$, $49 = 141.75377796 - 129.2343267\cos C$. Isolate for the trig function next and get $0.7177178102 = \cos C$ (this is why you don't *have* to memorize the

formulas for the angles!). Inverse cosine both sides of the equation to get $C \approx 44.1°$. Now it's easy to find B because the triangle's angles must total 180°. In this case, $B \approx 95.9°$.

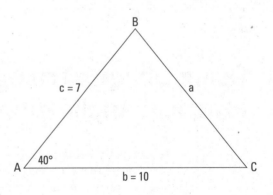

9 Solve the triangle if $C = 120°$, $a = 6$, and $b = 10$.

 10 Solve the triangle if $A = 70°$, $b = 6$, and $c = 7$.

 11 Solve the triangle if $a = 9$, $b = 5$, and $c = 7$.

12 Solve the triangle if $a = 7.3$, $b = 9.9$, and $c = 16$.

Using Oblique Triangles to Solve Practical Applications

Trigonometry word problems — these are great!. When you draw out a picture of the situation, you discover that each and every problem at this time has a triangle in it. The problems featured in this section aren't right triangles. They're *oblique triangles,* where you're looking for one missing piece of information (as opposed to looking for all the angles and sides, like you do in the previous three sections). That means less work for you! Each situation requires you to use the Law of Sines or the Law of Cosines exactly once to solve for the missing information.

Q. A plane flies for 300 miles in a straight line, makes a 45° turn, and continues for 700 miles. How far is it from its starting point?

EXAMPLE

A. Approximately 936.47 miles. First, draw out a picture like the one shown here. Notice that S is used for the starting point, T for the turning point, and E for the ending point.

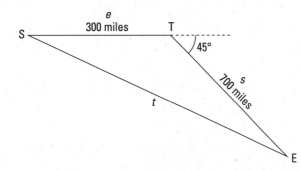

Now that you have the picture, you can figure out whether you need to use the Law of Sines or the Law of Cosines. Because the information given is SAS, you use a Law of Cosines, using the different variables from the picture. You're solving for the value of t, the length of the side opposite angle T. You have $e = 300$, $s = 700$, and angle T is $180° - 45° = 135°$. Plugging into $t^2 = s^2 + e^2 - 2(s)(e)\cos T$, you have $t^2 = 700^2 + 300^2 - 2(700)(300)\cos 135°$. Using your calculator, you get that $t^2 \approx 876,984.8481$ and $t \approx 936.47$.

Q. Two fire towers are exactly 5 miles apart in a forest. Personnel in the towers both spot a forest fire, one at an angle of 30° and the other at an angle of 42°. Which tower is closer?

EXAMPLE

A. The western fire tower is closer. In this case, you need to find two missing pieces of information. You have to find how far both towers are from the fire in order to know which one is closer. First, draw out a figure like the one that follows.

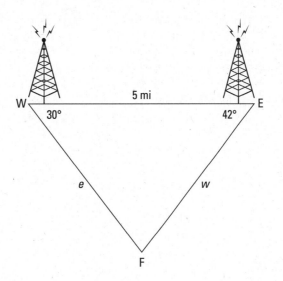

Use W for the western, E for the eastern tower, and F for the fire itself. You have a classic case of ASA, so you can use the Law of Sines this time. Knowing two of the angles makes it possible to find the third one easily: F = 108°. Now that you have the third angle, you can use the Law of Sines to form the needed proportions. You have $\frac{w}{\sin W} = \frac{e}{\sin E} = \frac{f}{\sin F}$. Filling in the known values, $\frac{w}{\sin 30°} = \frac{5}{\sin 108°}$.

Using the first and third ratios and solving for o, $\frac{w}{\sin 30°} = \frac{e}{\sin 42°} = \frac{5}{\sin 108°}$, you have $w = \frac{5\sin 30°}{\sin 108°}$, or $w \approx 2.6$ miles. Now set up another proportion to solve for e: $\frac{e}{\sin 42°} = \frac{5}{\sin 108°}$, which means that $e = \frac{5\sin 42°}{\sin 108°}$, or $t \approx 3.5$ miles.

The western tower is 2.6 miles from the fire, and the eastern tower is 3.5 miles from the fire.

---

13 Two trains leave a station at the same time on different tracks that have an angle of 100° between them. If the first train is a passenger train that travels 90 miles per hour and the second train is a cargo train that can travel only 50 miles per hour, how far apart are the two trains after three hours?

14 A radio tower is built on top of a hill. The hill makes an angle of 15° with the ground. The tower is 200 feet tall and located 150 feet from the bottom of the hill. If a wire is to connect the top of the tower with the bottom of the hill, how long does the wire need to be?

15 A surveyor stands on one side of a river looking at a flagpole on an island at an angle of 85°. She then walks in a straight line for 100 meters, turns, and looks back at the same flagpole at an angle of 40°. Find the distance from her first location to the flagpole.

16 Two scientists stand 350 feet apart, both looking at the same tree somewhere in between them. The first scientist measures an angle of 44° from the ground to the top of the tree. The second scientist measures an angle of 63° from the ground to the top of the tree. How tall is the tree?

Figuring Area

If you go in a slightly different direction in the proof of the Law of Sines, you discover a handy formula to find the area of an oblique triangle if you know two sides and the angle between them, as shown in Figure 10-1. The area of the triangle formed is $A = \frac{1}{2}ab\sin C$.

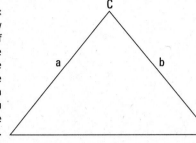

FIGURE 10-1:
If you know two sides of a triangle and the angle between them, you can calculate the area.

Note: The letters in the formula can change. The area of the triangle is *always* one-half the product of the two sides and the sine of the angle between them.

TECHNICAL STUFF

You can also find the area when you know all three sides (a, b, and c) by using what's called Heron's Formula. It says that the area of a triangle is

$$A = \sqrt{s(s-a)(s-b)(s-c)}, \; s = \frac{a+b+c}{2}$$

The variable s is called the semi-perimeter, or half of the triangle's perimeter.

EXAMPLE

Q. Find the area of the triangle where $b = 4$, $c = 7$, and $A = 36°$.

A. The area is about 8.23 square units. When you have the two sides and the angle between them, you plug the given information into the formula to solve for the area. In this case,

$$A = \frac{1}{2}(4)(7)\sin 36° \approx 8.23.$$

---

 17 Find the area of the triangle where $a = 7$, $c = 17$, and $B = 68°$.

18 Find the area of the triangle on the coordinate plane with vertices at (–5, 2), (5, 6), and (4, 0).

Answers to Problems on Solving Triangles

This section contains the answers for the practice problems presented in this chapter. You'll also find explanations on solving the particular types of problems.

(1) Solve the triangle if $B = 46°$, $C = 62°$, and $a = 21$. The answer is $A = 72°$, $b = 15.9$, and $c = 19.5$.

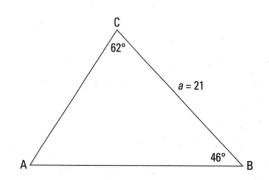

This one is ASA, so you use the Law of Sines to solve it. Because you already know two angles, begin by finding the third: $A = 180° - (46° + 62°) = 72°$. Now set up proportions from the Law of Sines to solve for the missing sides. From $\dfrac{21}{\sin 72°} = \dfrac{b}{\sin 46°}$, you have $b = \sin 46° \cdot \dfrac{21}{\sin 72°} \approx 15.9$.

And from $\dfrac{21}{\sin 72°} = \dfrac{c}{\sin 62°}$, you get that $c \approx 19.5$.

(2) Solve the triangle if $A = 19°$, $C = 100°$, and $b = 4.4$. The answer is $B = 61°$, $a \approx 1.64$, and $c \approx 4.95$.

Draw out the figure first. It's ASA, which means you use the Law of Sines to solve. Find the missing angle first: $B = 180° - (19° + 100°) = 61°$. Now set up the first proportion to solve for a:

$\dfrac{a}{\sin 19°} = \dfrac{4.4}{\sin 61°}$; $a = \sin 19° \cdot \dfrac{4.4}{\sin 61°} \approx 1.64$. Set up another proportion to solve for c:

$\dfrac{c}{\sin 100°} = \dfrac{4.4}{\sin 61°}$; $c = \sin 100° \cdot \dfrac{4.4}{\sin 61°} \approx 4.95$.

(3) Solve the triangle if $A = 49°$, $B = 21°$, and $a = 5$. The answer is $C = 110°$, $b \approx 2.37$, and $c \approx 6.23$.

This AAS case uses the Law of Sines. The missing angle is $C = 180° - (49° + 21°) = 110°$. The first proportion is $\dfrac{5}{\sin 49°} = \dfrac{b}{\sin 21°}$, which gets you $b = \sin 21° \cdot \dfrac{5}{\sin 49°} \approx 2.37$. The second proportion

is $\dfrac{5}{\sin 49°} = \dfrac{c}{\sin 110°}$, which gives you $c = \sin 110° \cdot \dfrac{5}{\sin 49°} \approx 6.23$.

(4) Solve the triangle if $A = 110°$, $C = 56°$, and $a = 8$. The answer is $B = 14°$, $b \approx 2.06$, and $c \approx 7.06$.

This one involves AAS, so you use the Law of Sines to solve it. First, the missing angle $B = 180° - (110° + 56°) = 14°$. Now set up the proportion, $\dfrac{8}{\sin 110°} = \dfrac{c}{\sin 56°}$, to get that

$c = \sin 56° \cdot \dfrac{8}{\sin 110°} \approx 7.06$. Set up another proportion, $\dfrac{8}{\sin 110°} = \dfrac{b}{\sin 14°}$, to get that

$b = \sin 14° \cdot \dfrac{8}{\sin 110°} \approx 2.06$.

(5) Solve the triangle if $b = 8$, $c = 14$, and $C = 37°$. The answer is A ≈ 122.9°, B ≈ 20.1°, and $a ≈ 19.53$.

Notice that when you draw this one that it's the SSA situation, or the ambiguous case. Always assume there are two answers when you're dealing with these types of problems, until you find out otherwise. Set up the proportion $\dfrac{14}{\sin 37°} = \dfrac{8}{\sin B}$. This means that $14 \cdot \sin B = 8 \sin 37°$, or $\sin B = \dfrac{8 \sin 37°}{14} ≈ 0.3438942989$. Use inverse sine to get that $B_1 ≈ 20.1°$. This is the first quadrant answer. The second quadrant has a second answer: $B_2 ≈ 180° - 20.1° = 159.9°$.

However, if you look closely, you notice that you start off with $C = 37°$. If you add $159.9° + 37°$ you get over 180°, so you throw this second solution away. Only one triangle satisfies the conditions given. Now that you know $C = 37°$ and (the one and only) $B ≈ 20.1°$, it's easy as pi (Get it? Pi!) to find $A = 122.9°$. Set up another proportion, $\dfrac{14}{\sin 37°} = \dfrac{a}{\sin 122.9°}$, which means that $a = \sin 122.9° \cdot \dfrac{14}{\sin 37°} ≈ 19.53$.

(6) Solve the triangle if $b = 5$, $c = 12$, and $B = 20°$. The answer is $A_1 ≈ 104.8°$, $C_1 ≈ 55.2°$, and $a_1 ≈ 14.13$; or, $A_2 ≈ 35.2°$, $C_2 ≈ 124.8°$, and $a_2 ≈ 8.42$.

Two solutions! How did that happen? Start at the beginning (a *very* good place to start) and use the Law of Sines to set up the proportion $\dfrac{5}{\sin 20°} = \dfrac{12}{\sin C}$. By cross-multiplying, you get the equation $5 \cdot \sin C = 12 \sin 20°$. Solve for $\sin C$ by dividing the 5: $\sin C = \dfrac{12 \sin 20°}{5} ≈ 0.820848344$.

Use the inverse sine function to discover that $C_1 ≈ 55.2°$. The second quadrant answer is $C_2 = 180° - 55.2° = 124.8°$. If you add 20° to *each* of these answers, you discover that it's possible to make a triangle in both cases (because you haven't exceeded 180°). This sends you on two different paths for two different triangles. The following shows the separate computations.

If $C_1 ≈ 55.2°$, then $A_1 ≈ 104.8°$. Next, set up the proportion $\dfrac{5}{\sin 20°} = \dfrac{a_1}{\sin 104.8°}$. This gives you $a_1 = \sin 104.8° \cdot \dfrac{5}{\sin 20°} ≈ 14.13$.

If $C_2 = 124.8°$, then $A_2 ≈ 35.2°$. Set up another proportion, $\dfrac{5}{\sin 20°} = \dfrac{a_2}{\sin 35.2°}$, to then get that $a_2 = \sin 35.2° \cdot \dfrac{5}{\sin 20°} ≈ 8.42$.

(7) Solve the triangle if $a = 10$, $c = 24$, and $A = 102°$. The answer is no solution.

When draw this one, you see that you have an ambiguous SSA case. Set up the proportion $\dfrac{10}{\sin 102°} = \dfrac{24}{\sin C°}$ using the Law of Sines. Cross-multiply to get $10 \cdot \sin C° = 24 \sin 102°$, and then divide the 10 from both sides to get $\sin C° = \dfrac{24 \sin 102°}{10} ≈ 2.35$. That's when the alarms go off. Sine can't have a value bigger than 1, so there's no solution.

(8) Solve the triangle if $b = 8$, $c = 16$, and $B = 30°$. The answer is $A = 60°$, $C = 90°$, and $a ≈ 13.86$.

Starting with the proportion $\dfrac{8}{\sin 30°} = \dfrac{16}{\sin C°}$ you cross-multiply to get $8 \sin C° = 16 \cdot \sin 30°$.

Dividing, $\sin C° = \dfrac{16 \cdot \sin 30°}{8} = 1$. You don't even need a calculator for this one. An angle of 90° has a sine equaling 1. So $C = 90°$, leaving 60° for angle A. You can use the side proportions for a

30-60-90 degree triangle or just go to the proportion $\dfrac{8}{\sin 30°} = \dfrac{a}{\sin 60°}$, which gives you $a = \sin 60° \cdot \dfrac{8}{\sin 30°} \approx 13.85640646$. The length 13.86 is the approximate value of $8\sqrt{3}$, which finishes the special 30-60-90 right triangle proportions of $\dfrac{1}{2}n, \dfrac{\sqrt{3}}{2}n, n$.

9 Solve the triangle if C = 120°, $a = 6$, and $b = 10$. The answer is $c = 14$, A ≈ 21.8°, and B ≈ 38.2°.

You see that this is a Law of Cosines problem when you draw the triangle (SAS). Find c first: $c^2 = a^2 + b^2 - 2ab \cos C$. Plug in what you know: $c^2 = 6^2 + 10^2 - 2(6)(10)\cos 120°$. Plug this right into your calculator to get that $c^2 = 196$, or $c = 14$. Now find A: $a^2 = b^2 + c^2 - 2bc \cos A$. Plug in what you know: $6^2 = 10^2 + 14^2 - 2(10)(14)\cos A$. Simplify: $36 = 296 - 280 \cos A$. Solve for $\cos A$: $-260 = -280 \cos A$ gives you $\dfrac{260}{280} = \cos A$ or $\cos A \approx 0.9285714286$. Now use inverse cosine to get A ≈ 21.8°, and then use the fact that the sum of the angles of a triangle is 180° to figure out that B ≈ 38.2°.

10 Solve the triangle if A = 70°, $b = 6$, and $c = 7$. The answer is $a \approx 7.50$, B ≈ 48.7°, and C ≈ 61.3°.

By plugging what you know into the Law of Cosines, $a^2 = b^2 + c^2 - 2bc \cos A$, you get $a^2 = 6^2 + 7^2 - 2(6)(7)\cos 70°$. This simplifies to $a^2 \approx 56.27030796$, or $a \approx 7.50$.

Now switch the substituting in the second law: $b^2 = a^2 + c^2 - 2ac \cos B$; $6^2 = 7.50^2 + 7^2 - 2(7.50)(7)\cos B$. Simplify: $36 = 105.27 - 105 \cos B$. Solving for $\cos B$: $-69.27 = -105 \cos B$, $\dfrac{69.27}{105} = \cos B$, resulting in $\cos B \approx 0.6597142857$. Using the inverse operation, you get that B ≈ 48.7°. From there you can figure out that C ≈ 61.3°.

11 Solve the triangle if $a = 9$, $b = 5$, and $c = 7$. The answer is A ≈ 95.7°, B ≈ 33.6°, and C ≈ 50.7°.

You're solving an SSS triangle using the Law of Cosines, so, to find A: $a^2 = b^2 + c^2 - 2bc \cos A$, $81^2 = 5^2 + 7^2 - 2(5)(7)\cos A$, $\cos A = \dfrac{7}{-70} = -0.1$, A ≈ 95.7°.

And again to find B: $b^2 = a^2 + c^2 - 2ac \cos B$, $5^2 = 9^2 + 7^2 - 2(9)(7)\cos B$, $\cos B = \dfrac{-105}{-126} \approx 0.83333333$, B ≈ 33.6°.

Last, but certainly not least, C = 180° − (95.7° + 33.6°) = 50.7°.

12 Solve the triangle if $a = 7.3$, $b = 9.9$, and $c = 16$. The answer is A ≈ 18.3°, B ≈ 25.3°, and C ≈ 136.4°.

To find A: $a^2 = b^2 + c^2 - 2bc \cos A$, $7.3^2 = 9.9^2 + 16^2 - 2(9.9)(16)\cos A$, $\cos A = \dfrac{-300.72}{-316.8} \approx 0.9492424242$, A ≈ 95.7°.

And to find B: $b^2 = a^2 + c^2 - 2ac \cos B$, $9.9^2 = 7.3^2 + 16^2 - 2(7.3)(16)\cos B$, $\cos B = \dfrac{-211.28}{-233.6} \approx 0.9044520548$, B ≈ 25.3°.

And finally, C ≈ 136.4°.

(13) Two trains leave a station at the same time on different tracks that have an angle of 100° between them. If the first train is a passenger train that travels 90 miles per hour and the second train is a cargo train that can travel only 50 miles per hour, how far apart are the two trains after three hours? The answer is approximately 330.86 miles apart.

The two trains depart (F and S) from the same station (T) in a picture like this one:

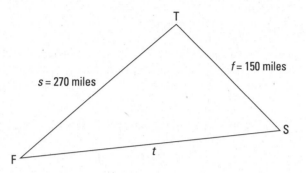

You have to use the Law of Cosines to solve for how far apart the two trains are, *t*.

To find how far the trains have traveled in three hours, use $d = rt$. For the first train you get $d = 90 \cdot 3 = 270$ miles; for the second train, $d = 50 \cdot 3 = 150$ miles. Plug these values into the equation: $t^2 = f^2 + s^2 - 2fs\cos T$. Plugging in values: $t^2 = 150^2 + 270^2 - 2(150)(270)\cos100°$. This goes right into your calculator to give you $t^2 = 109,465.50$, or *t* is about 330.86 miles.

(14) A radio tower is built on top of a hill. The hill makes an angle of 15° with the ground. The tower is 200 feet tall and located 150 feet from the bottom of the hill. If a wire is to connect the top of the tower with the bottom of the hill, how long does the wire need to be? The answer is about 279.3 feet long.

This time the picture is:

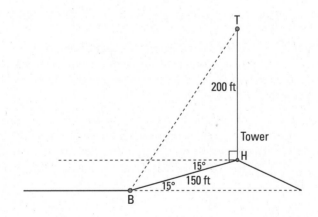

To find the measure of ∠H in the picture, you add a horizontal line that's parallel to the ground. Then, using the facts that alternate interior angles are congruent and that the tower has to be completely vertical (or else we have a leaning tower), you know that H = 15° + 90° = 105°.

Now, jump in with the Law of Cosines:

$h^2 = t^2 + b^2 - 2tb\cos H$.

Using the known values, $h^2 = 150^2 + 200^2 - 2(150)(200)\cos 105°$.

$h^2 = 78,029.14$ giving you $h \approx 279.3$ feet.

15) A surveyor stands on one side of a river looking at a flagpole on an island at an angle of 85°. She then walks in a straight line for 100 meters, turns, and looks back at the same flagpole at an angle of 40°. Find the distance from her first location to the flagpole. The answer is 78.5 meters.

Looking down on the surveyor and the flagpole, here's the picture you use to solve this problem:

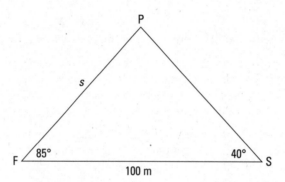

Because you already have two angles, you can find that P = 55° and use the Law of Sines:

$\dfrac{100}{\sin 55°} = \dfrac{s}{\sin 40°}$. Solving for s, you have $s = \sin 40° \cdot \dfrac{100}{\sin 55°} \approx 78.5$ meters.

16) Two scientists stand 350 feet apart, both looking at the same tree somewhere between them. The first scientist measures an angle of 44° from the ground to the top of the tree. The second scientist measures an angle of 63° from the ground to the top of the tree. How tall is the tree? The answer is 226.53 feet tall.

This problem involves two different triangles: the large triangle with the angles given, and a second triangle drawn with a segment through the tree as one side. You have to figure out the distance from either scientist to the top of the tree (FT or TS in the following figure) to determine how tall the tree (TB) really is. Here's a drawing of the two scientists and the tree between them:

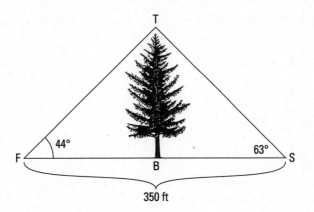

Knowing two angles gets you the third one: $T = 73°$. To determine the length of FT, use the Law of Sines: $\dfrac{350}{\sin 73°} = \dfrac{FT}{\sin 63°}$. Solving for FT, $FT = \sin 63° \cdot \dfrac{350}{\sin 73°} \approx 326.10$ feet.

Now, assuming the tree grows straight up, you have a right triangle FBT in which you know one angle and one side. That means you can to go back to SOHCAHTOA (see Chapter 6 for more information) to solve for the side, TB. Knowing that you have the hypotenuse and are looking for the opposite side, you can use the sine function and $\sin 44° = \dfrac{TB}{FT} = \dfrac{TB}{326.10}$. Solving for TB: $TB = 326.10 \cdot \sin 44° \approx 226.53$ feet tall. That's one big tree!

(17) Find the area of the triangle where $a = 7$, $c = 17$, and $B = 68°$. The answer is about 55.17 square units.

Knowing two sides and the angle between them allows you to use the area formula $A = \dfrac{1}{2} ab \sin C$. In this case $A = \dfrac{1}{2}(7)(17)\sin 68° \approx 55.17$.

(18) Find the area of the triangle on the coordinate plane with vertices at $(-5, 2)$, $(5, 6)$, and $(4, 0)$. The answer is about 28 square units.

Since you can use the distance formula to find the length of all three sides, you can then use Heron's Formula, $A = \sqrt{s(s-a)(s-b)(s-c)}$, to compute the area. So you should start by drawing a picture:

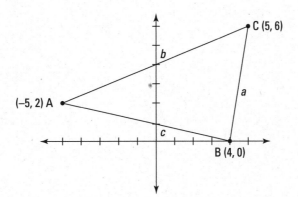

Find all three sides first (for a review of how to find the distance between two points, see Chapter 1):

$$AC = \sqrt{(5-(-5))^2 + (6-2)^2} = \sqrt{10^2 + 4^2} = \sqrt{116}$$

$$AB = \sqrt{(4-(-5))^2 + (0-2)^2} = \sqrt{9^2 + (-2)^2} = \sqrt{85}$$

$$BC = \sqrt{(4-5)^2 + (0-6)^2} = \sqrt{(-1)^2 + (-6)^2} = \sqrt{37}$$

Now that you've found the length of all three sides, use Heron's Formula to find the area.

First, the semi-perimeter is $s = \dfrac{\sqrt{116} + \sqrt{85} + \sqrt{37}}{2} \approx 13.0363$.

Using this in the formula, $A = \sqrt{13.0363(13.0363 - \sqrt{116})(13.0363 - \sqrt{85})(13.0363 - \sqrt{37})} \approx 28.05$.

4

Polar Coordinates, Cones, Solutions, Sequences, and Finding Your Limits

IN THIS PART . . .

This part begins with how to perform operations with and graph complex numbers. And then you see polar coordinates, a brand new way of graphing equations! Conic sections are also a great thing to graph, so they are covered in detail.

Moving on to the systems of equations you'll see solving linear and nonlinear equations, as well as working with matrices. Next, it's sequences and series. You'll see how to find any term in a sequence, how to calculate the sum of a sequence, and how to write the formula that determines a given sequence. Lastly, you'll find the topics that usually constitute the end of pre-calculus (and the beginning of calculus): limits and continuity.

Chapter **11**

Exploring Complex Numbers and Polar Coordinates

O nce upon a time, mathematicians delved into their imaginations and invented a whole new set of numbers. They needed these numbers so they could finish some math problems — problems where the square root of a negative number occurred.

Fields like engineering, electricity, and quantum physics all use imaginary numbers in their everyday applications. An *imaginary number* is basically the square root of a negative number. The *imaginary unit*, denoted i, is the solution to the equation $i^2 = -1$.

A *complex number* can be represented in the form $a + bi$, where a and b are real numbers and i denotes the imaginary unit. In the complex number $a + bi$, a is called the real part and b is called the imaginary part. Real numbers can be considered a subset of the complex numbers that have the form $a + 0i$. When a is zero, then $0 + bi$ is written as simply bi and is called a *pure imaginary number*.

And, just when it couldn't get any better, you also find in this chapter the process of graphing points and equations in a whole new way, called graphing polar coordinates.

Performing Operations with and Graphing Complex Numbers

Complex numbers in the form $a+bi$ can be graphed on a *complex coordinate plane*. Each complex number corresponds to a point (a, b) in the complex plane. The real axis is the line in the complex plane consisting of the numbers that have a zero imaginary part: $a+0i$. Every real number graphs to a unique point on the real axis. The imaginary axis is the line in the complex plane consisting of the numbers that have a zero real part: $0+bi$. Figure 11-1 shows several examples of points on the complex plane.

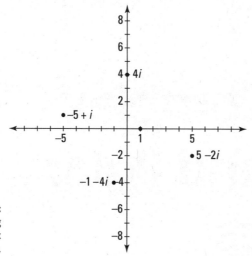

FIGURE 11-1:
Graphing
complex
numbers.

Adding and subtracting complex numbers is just another example of collecting like terms: You can add or subtract only real numbers, and you can add or subtract only imaginary numbers.

**TECHNICAL
STUFF**

When multiplying complex numbers, you FOIL the two binomials. All you have to do is remember that the imaginary unit is defined such that $i^2 = -1$, so any time you see i^2 in an expression, replace it with –1. When dealing with other powers of i, notice the pattern here:

$$i = i \qquad i^2 = -1 \qquad i^3 = -i \qquad i^4 = 1$$
$$i^5 = i \qquad i^6 = -1 \qquad i^7 = -i \qquad i^8 = 1$$

This continues in this manner forever, repeating in a cycle every fourth power. To find a larger power of i, rather than counting forever, realize that the pattern repeats. For example, to find i^{243}, divide 4 into 243 and you get 60 with a remainder of 3. The pattern will repeat 60 times and then you'll have 3 left over, so $i^{243} = i^{240} \cdot i^3 = 1 \cdot i^3$ which is $-i$.

The *conjugate* of a complex number $a+bi$ is $a-bi$, and vice versa. When you multiply two complex numbers that are conjugates of each other, you end up with a pure real number:

$$(a+bi)(a-bi) = a^2 - abi + abi - b^2 i^2$$

Combining like terms and replacing i^2 with –1: $= a^2 - b^2(-1) = a^2 + b^2$

Remember that absolute value bars enclosing a real number represent distance. In the case of a complex number, $|a + bi|$ represents the distance from the point to the origin. This distance is always the same as the length of the hypotenuse of the right triangle drawn when connecting the point to the x- and y-axes.

When dividing complex numbers, you multiply numerator and denominator by the conjugate. If the square root of a number is involved, then you'll be rationalizing the denominator.

In general, a division problem involving complex numbers looks like this:

$$\frac{a+bi}{c+di} = \frac{a+bi}{c+di} \cdot \frac{c-di}{c-di} = \frac{ac-adi+bci-bdi^2}{c^2-cdi+cdi-d^2i^2} = \frac{ac-adi+bci-bd(-1)}{c^2-d^2(-1)} = \frac{ac-adi+bci+bd}{c^2+d^2}$$

EXAMPLE

Q. Perform the indicated operations: $(3-4i)+(-2+5i)^2$.

A. $-18-24i$. Follow the order of operations and square the second binomial before combining the result with the first term. $(-2+5i)^2 = (-2+5i)(-2+5i) = 4-10i-10i+25i^2 = 4-20i+25(-1) = -21-20i$. Combine the like terms: $(3-4i)+(-21-20i) = -18-24i$.

Q. Perform the indicated operation: $\dfrac{6}{2-9i}$.

A. $\dfrac{12+54i}{85}$. The conjugate of the denominator is $2+9i$. Multiply this times the numerator and denominator of the fraction: $\dfrac{6}{2-9i} \cdot \dfrac{2+9i}{2+9i}$. Multiplying:
$\dfrac{6(2+9i)}{(2-9i)(2+9i)} = \dfrac{12+54i}{4-81i^2} = \dfrac{12+54i}{4+81}$.
This simplifies to the answer $\dfrac{12+54i}{85}$.

Q. Graph $4-6i$ on the complex coordinate plane.

A. See the following graph. Go to the right 4 units on the real axis and 6 units down on the imaginary axis and place a point. Easy!

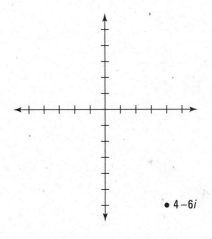

$\bullet\ 4-6i$

Q. Find real values x and y such that $3x + 4yi = 6 - 2i$.

A. $x = 2$, $y = -\dfrac{1}{2}$. The real parts of both sides of this equation must equal each other: $3x = 6$. Divide both sides by 3 to get the solution $x = 2$. Furthermore, the imaginary parts must also equal each other: $4y = -2$, so that when you divide both sides by 4 and reduce, you get $y = -\dfrac{1}{2}$.

1 Plot the point $4 - 3i$ on the complex plane.

2 Find $|3 - 4i|$.

 Find i^{22}.

 Solve the equation $5x^2 - 2x + 3 = 0$.

Round a Pole: Graphing Polar Coordinates

Up until now, your graphing experiences may have been limited to the *rectangular coordinate system.* The rectangular coordinate system gets that name because it's based on two number lines perpendicular to each other. It's now time to take that concept further and introduce *polar coordinates.*

In polar coordinates, every point is located around a central point, called the *pole,* and is named (r, θ). r is the radius, and θ is the angle formed between the polar axis (think of it as what *used* to be the positive *x*-axis) and the segment connecting the point to the pole (what *used* to be the origin).

In polar coordinates, angles are labeled in either degrees or radians (or both). If you need a recap on radians, see Chapter 6. Figure 11-2 shows the polar coordinate plane.

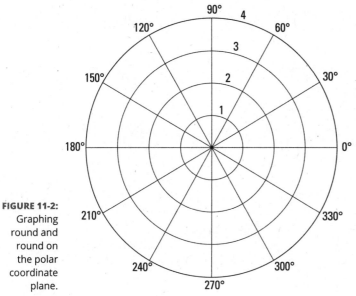

FIGURE 11-2:
Graphing round and round on the polar coordinate plane.

REMEMBER Notice that a point on the polar coordinate plane can have more than one name. Because you're moving in a circle, you can always add or subtract 2π to any angle and end up at the same point. This is an important concept when graphing equations in polar forms, so this section will cover it well.

TIP When both the radius and the angle are positive, the angle moves in a counterclockwise direction. If the radius is positive and the angle is negative, the point moves in a clockwise direction. If the radius is negative and the angle is positive, find the point where both are positive first and then reflect that point across the pole. If both the radius and the angle are negative, find the point where the radius is positive and the angle is negative and then reflect that across the pole.

Q. What's the polar coordinate of Point P in the following figure?

EXAMPLE

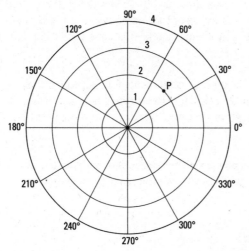

A. $(2, 45°)$ or $\left(2, \dfrac{\pi}{4}\right)$. First determine the radius by noticing that the point is 2 units away from the pole. Also notice that an angle forms when you connect the given point to the pole and that the polar axis is halfway between 30° and 60° or a 45°.

Q. Name two sets of coordinates that determine the same point in the preceding question.

A. Possible answers include $\left(2, \dfrac{9\pi}{4}\right)$, $\left(2, \dfrac{17\pi}{4}\right)$ and $\left(2, -\dfrac{7\pi}{4}\right)$. You add 2π to the angle twice and subtract it once to get these three angles. It's all about finding the common denominator at that point: $\dfrac{\pi}{4} + 2\pi = \dfrac{\pi}{4} + \dfrac{8\pi}{4} = \dfrac{9\pi}{4}$, for the first answer. Then take that answer and add 2π to it to get the next one. You can do this forever and still not list all the possibilities.

---

5 Graph $\left(4, -\dfrac{5\pi}{3}\right)$.

6 Graph $\left(-5, \dfrac{\pi}{2}\right)$.

7 Name two other polar coordinates for the point in Question 5: one with a negative angle and one with a positive angle.

8 Name two other polar coordinates for the point in Question 6: one with a negative angle and one with a positive radius.

Changing to and from Polar

You can use both polar and rectangular coordinates to name the same point on the coordinate plane. Sometimes it's easier to write an equation in one form than the other, so this should familiarize you with the choices and how to switch from one to another. Figure 11-3 shows how to determine the relationship between these two not-so-different methods.

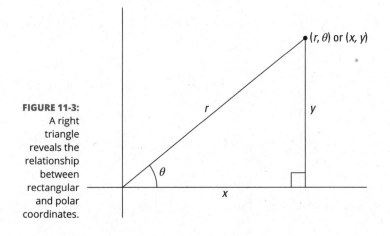

FIGURE 11-3: A right triangle reveals the relationship between rectangular and polar coordinates.

Some right triangle trigonometry and the Pythagorean Theorem:

TECHNICAL STUFF

» $\sin\theta = \dfrac{y}{r}$ $\leftrightarrow$ $y = r\sin\theta$

» $\cos\theta = \dfrac{x}{r}$ $\leftrightarrow$ $x = r\cos\theta$

» $\tan\theta = \dfrac{y}{x}$ $\leftrightarrow$ $\theta = \tan^{-1}\left(\dfrac{y}{x}\right)$

» $x^2 + y^2 = r^2$

EXAMPLE

Q. Rewrite the equation in rectangular form: $r = 2\sin\theta$.

A. $x^2 + y^2 = 2y$. First, convert the trig function to x, y, and r. If $r = 2\sin\theta$, then $r = 2\cdot\dfrac{y}{r}$. Then multiply both sides by r to get $r^2 = 2\cdot y$. Then you can replace r^2 from the Pythagorean Theorem and get $x^2 + y^2 = 2y$.

Q. Rewrite the equation $r = 2\csc\theta$ in rectangular form.

A. $y = 2$. First, realize that cosecant is the reciprocal of sine (see Chapter 6 for a refresher). If $\sin\theta = \dfrac{y}{r}$, then $\csc\theta = \dfrac{r}{y}$. Substitute this into $r = 2\csc\theta$ to get $r = 2\cdot\dfrac{r}{y}$.

Multiply both sides by y to get $ry = 2r$. Observe that the original equation implies that r is not zero, which allows you to cancel the r's here. This means that $y = 2$.

---

9 Rewrite $r = 3\sin\theta + 4\cos\theta$ in rectangular form.

10 Rewrite $y = 2x - 1$ in polar form.

11 Rewrite $3x - 5y = 10$ in polar form.

12 Rewrite $x^2 + y^2 = 16$ in polar form.

13 Convert the polar coordinates $\left(4, \dfrac{\pi}{6}\right)$ to rectangular coordinates.

 14 Convert the rectangular coordinates $(-1, 1)$ to polar coordinates.

Graphing Polar Equations

When given an equation in polar format and asked to graph it, you can always go with the plug-and-chug method: Pick values for θ from the unit circle that you know so well and find the corresponding value of r. Polar equations have various types of graphs, and it's easier to graph them if you have a general idea what they look like. Be sure to also see Chapter 12 (conic sections) for information on how to graph conic sections in polar coordinates.

Archimedean spiral

$r = a\theta$ gives a graph that forms a spiral. a is a constant that's multiplying the angle. If a is positive, the spiral moves in a counterclockwise direction, just like positive angles do. If a is negative, the spiral moves in a clockwise direction.

Cardioid

You may recognize the word *cardioid* if you've ever worked out and done your cardio. The word relates to the heart, and when you graph a cardioid, it does look like a heart, of sorts. Cardioids are written in the form $r = a(1 \pm \sin\theta)$ or $r = a(1 \pm \cos\theta)$. The cosine equations are hearts that point to the left or right, and the sine equations open up or down.

Rose

A rose by any other name is . . . a polar equation. If $r = a\sin b\theta$ or $r = a\cos b\theta$, the graphs look like flowers with petals. The number of petals is determined by b. If b is odd, then there are b (the same number of) petals. If b is even, there are $2b$ petals.

Circle

When $r = a\sin\theta$ or $r = a\cos\theta$, you end up with a circle with a diameter of a. Circles with cosine in them are centered on the x-axis, and circles with sine in them are centered on the y-axis. These are particular types of circles passing through the origin.

Lemniscate

A lemniscate makes a figure eight; that's the best way to remember it. $r^2 = \pm a^2 \sin 2\theta$ forms a figure eight between the axes, and $r^2 = \pm a^2 \cos 2\theta$ forms a figure eight that lies on one of the axes as a line of symmetry.

Limaçon

A cardioid is really a special type of limaçon, which is why they look similar to each other when you graph them. The familiar forms of limaçons are $r = a \pm b\sin\theta$ or $r = a \pm b\cos\theta$.

Q. Sketch the graph of $r = 1 + \sin\theta$.

A. See the following graph. This is a cardioid. If you're going to use a graphing calculator to graph polar equations, make sure your calculator is set to radians. If not, use the equation to find a few points to help you sketch in the curve. For this cardioid, if $\theta = 0$, then $r = 1 + \sin 0 = 1 + 0 = 1$. If $\theta = \frac{\pi}{2}$, then $r = 1 + \sin\frac{\pi}{2} = 1 + 1 = 2$. Keep going in this manner until you have enough points.

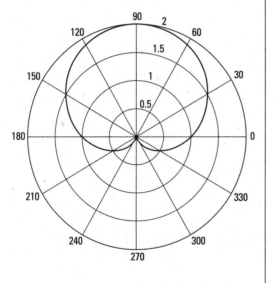

Q. Sketch the graph of $r = 3\theta$.

A. See the following graph. This is a rose with three petals because the coefficient on θ is odd. Plug and chug a few points, as shown in the following chart (only first quadrant values are shown here):

θ	$\cos 3\theta$	r
0	$\cos 3(0)$	1
$\frac{\pi}{6}$	$\cos\frac{\pi}{2}$	0
$\frac{\pi}{4}$	$\cos\frac{3\pi}{4}$	$-\frac{\sqrt{2}}{2}$
$\frac{\pi}{3}$	$\cos\pi$	-1
$\frac{\pi}{2}$	$\cos\frac{3\pi}{2}$	0

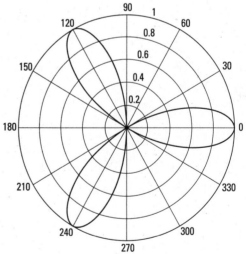

15 Sketch the graph of $r^2 = 9\cos 2\theta$.

16 Sketch the graph of $r = 2\theta$.

17 Sketch the graph of $r = 1 - 3\sin\theta$.

18 Sketch the graph of $r = 3 + 2\sin\theta$.

Answers to Problems on Complex Numbers and Polar Coordinates

Following are the answers to problems dealing with complex numbers and polar coordinates. You'll also find explanations about the answers.

(1) Plot the point $4 - 3i$ on the complex plane. See the following graph for the answer.

The real unit is 4 to the right and the imaginary unit is 3 down. This lands you at the point in the figure.

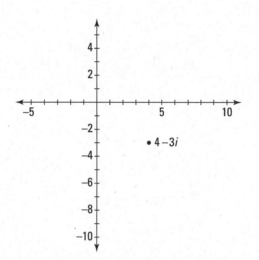

(2) Find $|3 - 4i|$. The answer is 5.

Remember that absolute value bars represent distance. In the case of complex numbers, they represent the distance from the point to the origin. This distance is always the same as the length of the hypotenuse of the right triangle drawn when connecting the point to the x- and y-axes.

$$d = \sqrt{3^2 + (-4)^2} = \sqrt{9 + 16} = \sqrt{25} = 5$$

(3) Find i^{22}. The answer is -1.

The pattern for the powers of i repeats every four times when its exponent increases consecutively. Divide 22 by 4 and get 5 with a remainder of 2. This means that $i^{22} = i^{20} \cdot i^2 = 1 \cdot i^2 = i^2 = -1$.

(4) Solve the equation $5x^2 - 2x + 3 = 0$. The answer is $x = \dfrac{1 \pm i\sqrt{14}}{5}$.

This quadratic equation can't be factored, so you have to resort to the quadratic formula to solve it: $x = \dfrac{-(-2) \pm \sqrt{(-2)^2 - 4(5)(3)}}{2(5)} = \dfrac{2 \pm \sqrt{4 - 60}}{10} = \dfrac{2 \pm \sqrt{-56}}{10} = \dfrac{2 \pm \sqrt{(-1)(4)(14)}}{10} = \dfrac{2 \pm 2i\sqrt{14}}{10}$.

This simplifies to $\dfrac{1 \pm i\sqrt{14}}{5}$. Written as complex numbers, you have $x = \dfrac{1}{5} + \dfrac{\sqrt{14}}{5}i$ and

$x = \dfrac{1}{5} - \dfrac{\sqrt{14}}{5}i$. If you've forgotten how to deal with the quadratic formula and/or how to simplify roots, see Chapter 4.

(5) Graph $\left(4, -\dfrac{5\pi}{3}\right)$. See the following graph for the answer.

The radius is 4 and the angle is negative, which moves in a clockwise direction and ends up in the first quadrant. See Chapter 6, page 124, for the equivalent radian and degree measures.

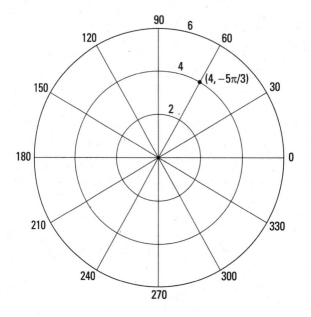

(6) Graph $\left(-5, \dfrac{\pi}{2}\right)$. See the following graph for the answer.

The radius for this one is negative, which reflects the point $\left(5, \dfrac{\pi}{2}\right)$ over the pole. See Chapter 6, page 124, for the equivalent radian and degree measures.

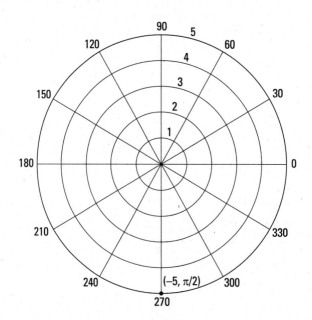

(7) Name two other polar coordinates for the point in Question 5: one with a negative angle and one with a positive angle. Possible answers include $\left(4,-\dfrac{11\pi}{3}\right)$ and $\left(4,\dfrac{\pi}{3}\right)$.

The radius isn't changing, so don't do anything to it. Subtract 2π from the angle to get the first answer. Add 2π to the angle to get the second answer.

(8) Name two other polar coordinates for the point in Question 6: one with a negative angle and one with a positive radius. Possible answers include $\left(-5,-\dfrac{3\pi}{2}\right)$ and $\left(5,\dfrac{3\pi}{2}\right)$.

To deal with the first situation, don't change the radius; just subtract 2π to get a negative angle: $-\dfrac{3\pi}{2}$. To change the radius to a positive 5, you have to change the angle. Because the point was down 5 originally, this angle is the same as $\dfrac{3\pi}{2}$.

(9) Rewrite $r = 3\sin\theta + 4\cos\theta$ in rectangular form. The answer is $x^2 + y^2 = 3y + 4x$.

First, change $\sin\theta$ to $\dfrac{y}{r}$ and $\cos\theta$ to $\dfrac{x}{r}$: $r = 3\cdot\dfrac{y}{r} + 4\cdot\dfrac{x}{r}$. Multiply each term by r to get rid of the fractions and get $r^2 = 3y + 4x$. Use the Pythagorean substitution and you get $x^2 + y^2 = 3y + 4x$.

(10) Rewrite $y = 2x - 1$ in polar form. The answer is $r = \dfrac{-1}{\sin\theta - 2\cos\theta}$.

Remember that $x = r\cos\theta$ and $y = r\sin\theta$. Make these substitutions first and get $r\sin\theta = 2r\cos\theta - 1$. Get all terms with r to one side first: $r\sin\theta - 2r\cos\theta = -1$. Factor out the common factor of r: $r(\sin\theta - 2\cos\theta) = -1$. Now solve for r by dividing: $r = \dfrac{-1}{\sin\theta - 2\cos\theta}$ or $r = \dfrac{1}{2\cos\theta - \sin\theta}$.

(11) Rewrite $3x - 5y = 10$ in polar form. The answer is $r = \dfrac{10}{3\cos\theta - 5\sin\theta}$.

Use the same substitutions as in question 10 to get $3r\cos\theta - 5r\sin\theta = 10$. Factor out the r and divide both sides by its multiplier: $r(3\cos\theta - 5\sin\theta) = 10$ becomes $r = \dfrac{10}{3\cos\theta - 5\sin\theta}$.

(12) Rewrite $x^2 + y^2 = 16$ in polar form. The answer is $r = 4$.

Use the Pythagorean substitution first: $x^2 + y^2 = r^2$, so $r^2 = 16$. Take the square root of both sides and get $r = 4$. Of course, $r = -4$ is also a way to write the equation for this circle.

(13) Convert the polar coordinate $\left(4,\dfrac{\pi}{6}\right)$ to rectangular coordinates. The answer is $\left(2\sqrt{3},2\right)$.

Since $x = r\cos\theta$, then $x = 4\cos\dfrac{\pi}{6} = 4\left(\dfrac{\sqrt{3}}{2}\right) = 2\sqrt{3}$.

Since $y = r\sin\theta$, then $y = 4\sin\dfrac{\pi}{6} = 4\left(\dfrac{1}{2}\right) = 2$.

(14) Convert the rectangular coordinate $(-1,1)$ to polar coordinates. The answer is $\left(\sqrt{2},-\dfrac{\pi}{4}\right)$.

Use the Pythagorean substitution first to find r: $r^2 = x^2 + y^2 = (-1)^2 + 1^2 = 2$ means that $r^2 = 2$, or $r = \sqrt{2}$. To find θ, use $\tan\theta = \dfrac{y}{x}$, which in this case, is $\tan\theta = \dfrac{1}{-1} = -1$. The tangent is -1 when $\theta = -\dfrac{\pi}{4}$.

(15) Sketch the graph of $r^2 = 9\cos 2\theta$. See the following graph for the answer.

This graph is a lemniscate. If $\theta = 0$, $r = 3$ or $r = -3$. If $\theta = \dfrac{\pi}{6}$, then $r \approx 2.12$ or $r \approx -2.12$, and so on. Use these points and others to sketch the graph.

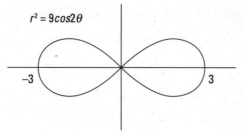

(16) Sketch the graph of $r = 2\theta$. See the following graph for the answer.

This is a spiral. If $\theta = 0$, $r = 0$; if $\theta = \dfrac{\pi}{6}$, $r \approx 1.047$, and so on.

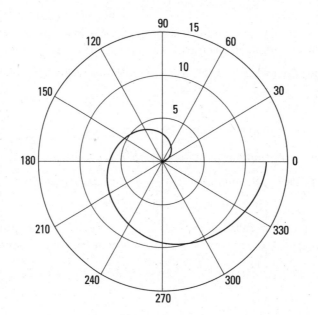

17 Sketch the graph of $r = 1 - 3\sin\theta$. See the following graph for the answer.

This is a limaçon. Plug and chug it with enough points to get the graph.

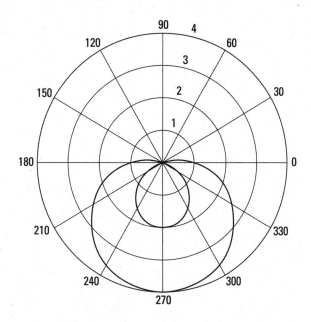

18 Sketch the graph of $r = 3 + 2\sin\theta$. See the following graph for the answer.

This is a cardioid. If you plug and chug it, you end up with the graph.

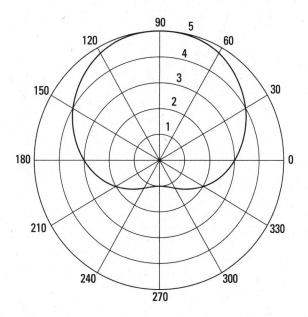

Chapter **12**

Conquering Conic Sections

Who doesn't love a good cone? Some are big fans of a double-dip ice cream cone, but maybe you're partial to the traffic cone because it keeps you safe on the road. Whatever type of cone is your favorite, for mathematicians the cone is the creative fuel for the fire of a whole bunch of ideas.

You see, about 2,200 years ago, some smart mathematician named Apollonius of Perga decided to stack two cones point to point. He sliced them in different directions and came up with four different *conic sections:* the *circle*, the *ellipse*, the *parabola*, and the *hyperbola*. Each conic section has its own equation and its own parts, which you need to determine in order to graph it. You may wish to graph a conic section, identify certain parts, or write its equation. The basis of *any* of these tasks is to be able to recognize what kind of conic section it is and write it in its own equation form.

When a conic section is already written in its standard form, you know most of the information about the parts of that particular conic. When a conic section isn't written in its standard form, you first have to write it that way. How do you do that? You find that information in this chapter.

A Quick Conic Review

A basic process used to change a conic section's equation into the standard form is *completing the square*. The steps of the process are as follows:

TECHNICAL STUFF

1. **Add/subtract any constant to the opposite side of the given equation, away from all the variables.**

2. **Factor the leading coefficient out of all terms in front of the set of parentheses.**

3. **Determine the result of dividing the linear coefficient by two.**

4. **Square the answer from Step 3 and add that inside the parentheses.**

   Don't forget that if you have a coefficient from Step 2, you must multiply the coefficient by the number you get in this step and add *that* to both sides.

5. **Factor the quadratic polynomial as a perfect square trinomial.**

This chapter is dedicated to the various mathematical conics. You see how to write each one in its form and how to graph it.

Going Round and Round with Circles

A circle is simple, but not so plain. You can do so much with a circle. The very tires you drive on are, after all, circular in shape. A circle has one point in the very middle called the *center*. All the points on the circle are the same distance from the center, and that distance is the *radius*.

When a circle is drawn centered at the origin of the coordinate plane, the equation that describes it is simple as well:

$$x^2 + y^2 = r^2$$

TECHNICAL STUFF

where r represents the circle's radius.

When the circle is moved around the coordinate plane with horizontal and/or vertical shifts, its equation looks like this:

$$(x-h)^2 + (y-k)^2 = r^2$$

The circle is centered at (h,k), h is the horizontal shift from the origin, k is the vertical shift from the origin, and r is still the radius. If a circle isn't written in this form, you'll still recognize the equation as one that may define a circle because the equation will have *both* an x^2 and a y^2, and the coefficients on both will be equal.

EXAMPLE

Q. What's the center and the radius of the circle $(x+3)^2 + (y-2)^2 = 16$?

A. The center is $(-3, 2)$; radius $r = 4$. This equation is already in the proper form of a circle, so finding its information is easy. The horizontal value with x is 3, so $h = -3$; the vertical shift is $+2$, because $k = 2$. Meanwhile, the other side of the equation gives you the radius squared. Setting $r^2 = 16$ gives you the radius $r = 4$.

EXAMPLE

Q. Graph $x^2 + y^2 - 6y + 2 = 0$.

A. See the following graph. Using completing the square, you first subtract the constant to move it to the other side: $x^2 + y^2 - 6y = -2$. Because the leading coefficients are 1, you don't have to factor anything out and can move onto the next step. Because the x variable doesn't have a linear term, you don't have to complete the square there. Notice, however, that the y variable does have a linear term, so you square half of the coefficient: $\left(\dfrac{-6}{2}\right)^2 = 9$. This value gets added to both sides: $x^2 + y^2 - 6y + 9 = -2 + 9$.

Factor the y's and the 9 as a perfect square trinomial and simplify on the right to get $x^2 + (y-3)^2 = 7$. You have the standard form of a circle. Its center is $(0, 3)$, and its radius is $r = \sqrt{7}$.

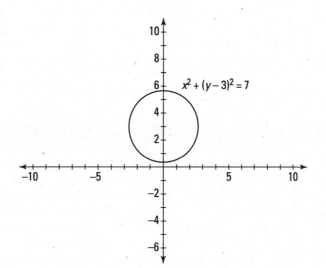

Find the center and radius of the circle $2x^2 + 2y^2 - 4x = 15$. Then graph the circle.

 Write the equation of the circle with the center $(-1, 4)$ if the circle passes through the point $(3, 1)$.

The Ups and Downs: Graphing Parabolas

When you graph a *quadratic polynomial* (see Chapter 4 for more information on this type of polynomial), you always get a parabola. Typically, up until this point in pre-calculus, the graphs of parabolas have been vertical: They open up or down. In conic sections, however, they can also open horizontally: to the left or to the right. You see these situations in the following sections.

Officially, a *parabola* is the set of all points on a plane that are the same distance from a given point (*focus*) and a given line (*directrix*). Each parabola can be folded exactly in half over a line called the *axis of symmetry*. The point where the axis of symmetry intersects the graph is called the *vertex*. This gives you what you might want to call the martini of conic sections: The parabola is the glass, the axis of symmetry is the stem, the directrix is the base, and the focus is the olive. Every good martini has all its parts, and every good parabola does, too. But don't be too shaken (or stirred) up! Figure 12-1 shows all the parts of a parabola.

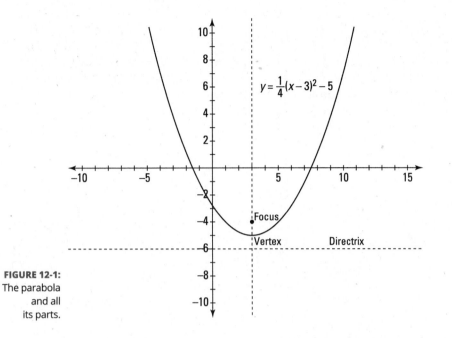

$$y = \frac{1}{4}(x-3)^2 - 5$$

Focus

Vertex   Directrix

FIGURE 12-1:
The parabola
and all
its parts.

Standing tall: Vertical parabolas

TECHNICAL STUFF

The equation of a vertical parabola is

$$y = a(x-h)^2 + k$$

where h is the horizontal shift from the origin, k is the vertical shift from the origin, and a is the vertical stretch or flattening. You know you're dealing with a vertical parabola when x is squared but y isn't. Each vertical parabola has the following parts:

» **Vertex:** (h, k)

» **Axis of symmetry:** $x = h$

» **Focus:** $\left(h, k + \dfrac{1}{4a}\right)$

» **Directrix:** $y = k - \dfrac{1}{4a}$

TIP

The vertex is usually the first point that you graph on a parabola. From there, know that the focus and the directrix are $\dfrac{1}{4a}$ away from it. One is above and one is below, depending on the value of a. If $a < 0$, the parabola opens down and the focus is below the vertex while the directrix is above. If $a > 0$, the parabola opens up, the focus is above the vertex, and the directrix is below.

Q. Graph the parabola $y = -2(x-1)^2 + 5$.

EXAMPLE **A.** See the following graph. Because this equation is already in the standard parabola form, you should be able to go right to graphing. The vertex is at $(1, 5)$. The vertical stretch is 2, and the graph is turned upside down. If you don't recognize where this information comes from, you can go back and read about it in Chapter 3.

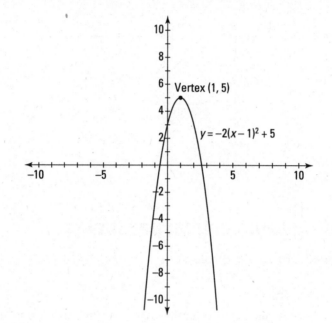

Q. State the vertex, axis of symmetry, focus, and directrix of $y = 3x^2 - 4x + 1$.

A. Vertex: $\left(\frac{2}{3}, -\frac{1}{3}\right)$; axis of symmetry: $x = \frac{2}{3}$; focus: $\left(\frac{2}{3}, -\frac{1}{4}\right)$; directrix: $y = -\frac{5}{12}$. That's an awful lot of fractions, but the process doesn't change. Start by subtracting 1 from both sides: $y - 1 = 3x^2 - 4x$. Then factor out the 3: $y - 1 = 3\left(x^2 - \frac{4}{3}x\right)$. Now complete the square and be sure to keep the equation balanced by multiplying the $\left(\frac{1}{2}\left(-\frac{4}{3}\right)\right)^2$ by 3 before add-

ing it to the left side: $y - 1 + \frac{4}{3} = 3\left(x^2 - \frac{4}{3}x + \frac{4}{9}\right)$. Simplify and factor: $y + \frac{1}{3} = 3\left(x - \frac{2}{3}\right)^2$.

Then solve for y to put the parabola in its proper form: $y = 3\left(x - \frac{2}{3}\right)^2 - \frac{1}{3}$. Next, use the

formulas to figure out all the parts — the vertex is $\left(\frac{2}{3}, -\frac{1}{3}\right)$, the axis of symmetry is

$x = \frac{2}{3}$, the focus is $\left(\frac{2}{3}, -\frac{1}{3} + \frac{1}{4(3)}\right) = \left(\frac{2}{3}, -\frac{1}{4}\right)$, and the directrix is $y = -\frac{1}{3} - \frac{1}{4(3)} = -\frac{5}{12}$.

3 What's the vertex of the parabola $y = -x^2 + 4x - 6$? Sketch the graph of this parabola.

4 Find the focus and the directrix of the parabola $y = 4x^2$.

Lying down on the job: Horizontal parabolas

TECHNICAL STUFF

The equation of a horizontal parabola is very similar to the vertical one I discuss in the preceding section. Here it is:

$$x = a(y - k)^2 + h$$

This is a horizontal parabola because y is squared but x isn't. Notice, also, that h and k are still there for the horizontal and vertical shifts, respectively, but that they've switched places. Because this parabola is horizontal, a also switches to become the horizontal stretch or flattening. A *horizontal stretch or flattening* does the same thing as its vertical counterpart, but it affects what the function does from left to right. A horizontal transformation where a is a fraction between 0 and 1 will flatten the curve, and a horizontal transformation where $a > 1$ will make it appear to steepen. All these numbers are positive, so the parabola opens to the right. When a is negative, the parabola does the same thing, but the graph is reflected in the opposite direction (so the parabola opens to the left).

Here are the parts of a horizontal parabola:

>> **Vertex:** (h, k)

>> **Axis of symmetry:** $y = k$

>> **Focus:** $\left(h + \dfrac{1}{4a}, k \right)$

>> **Directrix:** $x = h - \dfrac{1}{4a}$

Q. Graph the parabola if its equation is $x = y^2 - 6y$.

EXAMPLE

A. See the following figure. Complete the square to get this horizontal parabola in its form: $x + 9 = (y - 3)^2$ becomes $x = (y - 3)^2 - 9$. This means that the parabola's vertex is located at the point $(-9, 3)$ and the graph is symmetric about $y = 3$. There's no stretch or flattening ($a = 1$). Using these points and the symmetry of the parabola gives you the graph in the figure.

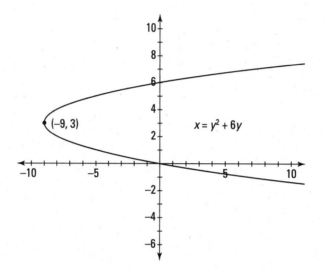

Q. Write the equation of the parabola whose vertex is $(-2, 1)$ if the focus is at the point $(-4, 1)$.

A. $x = -\dfrac{1}{8}(y - 1)^2 - 2$. Because the focus is to the left of the vertex, you know that the parabola is a horizontal one. Start with the general equation of a horizontal parabola: $x = a(y - k)^2 + h$. Then plug in the vertex values $(-2, 1)$: $-2 = a(y - 1)^2 + h$. Now, all you have to do is figure out what the value of a is. You know that the format to find the focus is $\left(h + \dfrac{1}{4a}, k\right)$. This tells you that $h + \dfrac{1}{4a} = -4$. You also know that h is -2, so substitute and get $-2 + \dfrac{1}{4a} = -4$. Solve to get $a = -\dfrac{1}{8}$. Finally, write the equation of the parabola: $x = -\dfrac{1}{8}(y - 1)^2 - 2$.

---

5 Sketch the graph of $x = 2(y - 4)^2$.

6 Determine whether this parabola opens left or right: $x = -y^2 - 7y + 3$.

The Fat and the Skinny: Graphing Ellipses

An *ellipse* is defined as the set of all points on a plane, such that the sum of the distances from any point on the curve to two fixed points, the foci, is a constant. Think of an ellipse as a circle that has gone flat, like an egg or oval mirror. An ellipse has its own unique parts and equations, depending on whether (you guessed it) it's horizontal or vertical. Here's what you need to know about any ellipse, whether it's horizontal or vertical:

>> The center is at the point (h, k).

>> The longer axis of symmetry is called the *major axis,* and the distance from the center to a point on the ellipse along the major axis, often called the *semi-major axis,* is represented by *a.* The points where the ellipse intersects this axis are called the *vertices.*

>> The shorter axis of symmetry is called the *minor axis,* and the distance from the center to a point on the ellipse along the minor axis, often called the *semi-minor axis,* is represented by *b.* The points where the ellipse intersects this axis are called the *co-vertices.*

>> This means that *a* is always greater than *b* in an ellipse.

>> You can find the foci of the ellipse along its major axis by using the equation $f^2 = a^2 - b^2$.

Figure 12-2 shows how all these pieces fall into place for a horizontal and a vertical ellipse.

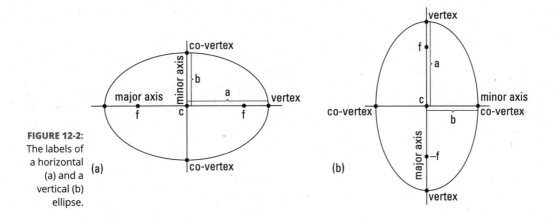

FIGURE 12-2: The labels of a horizontal (a) and a vertical (b) ellipse.

Short and fat: The horizontal ellipse

The equation of a horizontal ellipse is

$$\frac{(x-h)^2}{a^2} + \frac{(y-k)^2}{b^2} = 1.$$

TECHNICAL
STUFF

Notice that all the variables (except for f, the focus) make their appearance in the equation. Notice also that because $a > b$, for any ellipse, $a^2 > b^2$. The fact that the bigger number, a^2, is in the denominator of the x fraction tells you that the ellipse is horizontal. To graph any horizontal ellipse after it's written in this form, mark the center first. Then count out a units to the left and right and b units up and down. These four points help determine the ellipse's shape. The vertices are points found at $(h \pm a, k)$. The co-vertices are $(h, k \pm b)$. The two foci are $2f$ units apart along the major axis. As points, the foci are $(h \pm f, k)$.

Q. State the center, vertices, and foci of the ellipse $3x^2 - 18x + 5y^2 = 3$.

EXAMPLE

A. Center: $(3,0)$; vertices: $\left(3 \pm \sqrt{10}, 0\right)$; foci: $(5,0)$ and $(1,0)$. As usual, you need to complete the square to write this ellipse in its standard form. The constant is already on the right, so begin by factoring the x terms: $3\left(x^2 - 6x\right) + 5y^2 = 3$. Next, complete the square and balance the equation: $3\left(x^2 - 6x + 9\right) + 5y^2 = 3 + 27$. Factor the perfect square trinomial and simplify: $3(x-3)^2 + 5y^2 = 30$.

Divide every term by 30: $\dfrac{(x-3)^2}{10} + \dfrac{y^2}{6} = 1$.

This tells you the center is $(3,0)$. Then, since $a^2 = 10$, $a = \sqrt{10}$, and since $b^2 = 6$, $b = \sqrt{6}$. This gives you the vertices at $\left(3 \pm \sqrt{10}, 0\right)$. It also tells you the co-vertices are at $\left(3, \pm\sqrt{6}\right)$. Lastly, $f^2 = 10 - f^2 = 10 - 6 = 4$, so $f = 2$, and the foci are at $(5, 0)$ and $(1, 0)$.

Q. Sketch the graph of the ellipse

EXAMPLE

$$\frac{(x+2)^2}{25} + \frac{(y-1)^2}{16} = 1.$$

A. See the following figure. This ellipse is written in the proper form, so to graph it, all you have to do is identify its parts. The center is $(-2, 1)$. If $a^2 = 25$, then. This means your vertices are 5 units to the left and the right from the center, at $(-7, 1)$ and $(3, 1)$. If $b^2 = 16$, then. This means your co-vertices are 4 units above and below the center at $(-2, 5)$ and $(-2, -3)$.

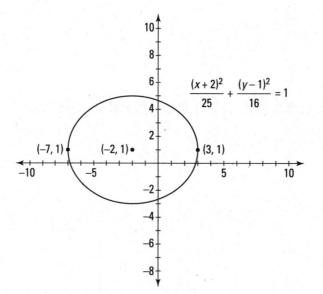

7 Sketch the graph of the ellipse
$4x^2 + 12y^2 - 8x - 24y = 0$.

8 Write the equation of the ellipse with vertices
at $(-1,1)$ and $(9,1)$ and foci at $\left(4 \pm \sqrt{21}, 1\right)$.

Tall and skinny: The vertical ellipse

TECHNICAL STUFF

The equation of a vertical ellipse is

$$\frac{(x-h)^2}{b^2} + \frac{(y-k)^2}{a^2} = 1.$$

This equation looks awfully familiar, doesn't it? The only difference between a horizontal ellipse and a vertical one is the location of a. When the bigger number is under x, it's a horizontal ellipse. When the bigger number is under y, it's a vertical ellipse. You graph this ellipse by marking the center, counting up and down a units to find the vertices, and then counting left and right b units to find the co-vertices. This means your vertices are at $(h, k \pm a)$ and your co-vertices are at $(h \pm b, k)$. Your foci move in the same direction as your vertices, so they're at $(h, k \pm f)$.

EXAMPLE

Q. Sketch the graph of the ellipse $81x^2 + 4y^2 = 324$. State the foci of this ellipse.

A. See the following graph; the foci are $\left(0, \pm\sqrt{77}\right)$.

Don't let the large numbers throw you. Just remember that your only goal is to write the equation in the proper form. The equation has no x variable or y variable to the first degree. That means you don't have to complete the square! Say what? All you need to do is get 1 on the right side of the equation by dividing everything by 324.

When you do, the equation reduces to $\frac{x^2}{4} + \frac{y^2}{81} = 1$, which also conveniently puts it in the form you want. This ellipse has its center at the origin $(0,0)$.

Since $a^2 = 81$, a moves up and down 9 units from the center, and while $b^2 = 4$, so b moves left and right 2 units and gives you the following graph. Lastly, $f^2 = 81 - 4 = 77$, so $f = \sqrt{77}$. This means your foci are at $\left(0, \pm\sqrt{77}\right)$.

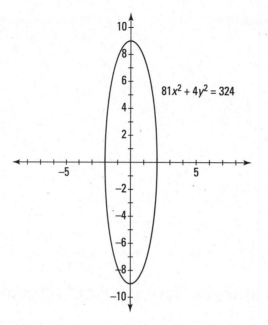

$81x^2 + 4y^2 = 324$

Q. Write the equation of the vertical ellipse with its center at $(-4, 1)$ if its major axis has a length of 10 and its minor axis has a length of 8.

A. $\dfrac{(x+4)^2}{16} + \dfrac{(y-1)^2}{25} = 1$. You have all the information you need to write the equation. The center is given to you as $(-4, 1)$. If the major axis has a length of 10, then $2a = 10$, or $a = 5$. Also, the minor axis has a length of 8, so $2b = 8$, or $b = 4$. Knowing that the ellipse is vertical tells you to put a^2 under the y term and b^2 under the x term. This gives you the equation $\dfrac{(x+4)^2}{16} + \dfrac{(y-1)^2}{25} = 1$.

9 Sketch the graph of the ellipse $\dfrac{(x-1)^2}{8} + \dfrac{(y+2)^2}{6} = 1$.

10 State the ordered pair for the vertices, the co-vertices, and the foci of the ellipse $(x+1)^2 + \dfrac{(y-4)^2}{16} = 1$.

No Caffeine Required: Graphing Hyperbolas

A *hyperbola* is the set of all points in the plane such that the difference of the distances from two fixed points (the foci) is a positive constant. Hyperbolas always come in two parts. Each one is a perfect mirror reflection of the other over their axis of symmetry. There are horizontal and vertical hyperbolas. Regardless of how the hyperbola opens, you always find the following parts:

>> The center is at the point (h,k).

>> As the curve moves farther and farther from the center, the graph on both sides gets closer and closer to two diagonal lines known as *asymptotes*. The equation of the hyperbola, regardless of whether it's horizontal or vertical, gives you two values: a and b. These help you draw a box, and when you draw the diagonals of this box, you find the asymptotes.

>> There are two axes of symmetry:

- The one passing through the *vertices* is called the *transverse axis*. The distance from the center along the transverse axis to the vertex is represented by a.

- The one perpendicular to the transverse axis through the center is called the *conjugate axis*. The distance along the conjugate axis from the center to the edge of the box that determines the asymptotes is represented by b.

- a and b have no relationship; a can be less than, greater than, or equal to b.

>> You can find the foci by using the equation $f^2 = a^2 + b^2$.

Figure 12-3 shows the parts of a vertical (opening upward and downward) hyperbola.

Horizontal hyperbolas

A horizontal hyperbola opens to the left and right; the curves are reflections of one another over the vertical or conjugate axis. The equation of a horizontal hyperbola is

$$\frac{(x-h)^2}{a^2} - \frac{(y-k)^2}{b^2} = 1.$$

TECHNICAL STUFF

This equation looks really similar to that of the horizontal ellipse. But if you look closely, you notice the subtraction sign between the two fractions. To begin graphing, identify a, which helps determine one edge of a box that you can use to find the hyperbola's asymptotes. The opposite corners of this imaginary box are two points of each of the asymptotes, so they can be used to draw those lines. The value of a is in the denominator of the x fraction, so it will be left and right from the center. The vertices are at $(h \pm a, k)$. The other edge of the box is found from b, under the y fraction. It moves up and down. The foci move in the same direction as a and can be found at $(h \pm f, k)$. Equations of the asymptotes of a horizontal hyperbola are given by $y = \pm \dfrac{b}{a}(x-h) + k$.

Q. Sketch the graph of $2x^2 - 3y^2 + 10x + 6y = \dfrac{41}{2}$.

EXAMPLE **A.** See the following figure. Put it in its standard form by completing the square. To do this, group the same variables together and factor out any common factors: $2x^2 + 10x - 3y^2 + 6y = \dfrac{41}{2}$ becomes $2(x^2 + 5x) - 3(y^2 - 2y) = \dfrac{41}{2}$. (Watch out for the negative sign when factoring.) Complete the square for each variable, keeping the equation balanced: $2\left(x^2 + 5x + \dfrac{25}{4}\right) - 3(y^2 - 2y + 1) = \dfrac{41}{2} + \dfrac{25}{2} - 3$. Factor and simplify

$2\left(x + \dfrac{5}{2}\right)^2 - 3(y-1)^2 = 30$. Last, divide every term by 30: $\dfrac{\left(x+\dfrac{5}{2}\right)^2}{15} - \dfrac{(y-1)^2}{10} = 1$. From the equation you know that the center is $\left(-\dfrac{5}{2}, 1\right)$, $a = \sqrt{15}$, and $b = \sqrt{10}$. So count left and right $\sqrt{15} \approx 3.9$ units from the center for the left and right sides of the box and count up and down $\sqrt{10} \approx 3.2$ from the center for the top and bottom of the box. Extend the sides until they meet, and draw the asymptotes through the opposite corners.

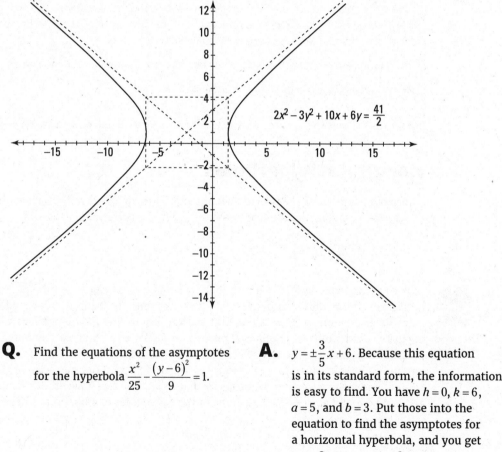

Q. Find the equations of the asymptotes for the hyperbola $\dfrac{x^2}{25} - \dfrac{(y-6)^2}{9} = 1$.

A. $y = \pm\dfrac{3}{5}x + 6$. Because this equation is in its standard form, the information is easy to find. You have $h = 0$, $k = 6$, $a = 5$, and $b = 3$. Put those into the equation to find the asymptotes for a horizontal hyperbola, and you get $y = \pm\dfrac{3}{5}(x - 0) + 6 = \pm\dfrac{3}{5}x + 6$.

11 Find the equation of the hyperbola that meets the following criteria: It has its center at $(4, 1)$, one of its vertices is at $(7, 1)$, and one of its asymptotes is $3y = 2x - 5$.

12 Sketch the graph of the equation $(2x - y)(x + 5y) - 9xy = 10$.

Vertical hyperbolas

The equation of a vertical hyperbola is

$$\frac{(y-k)^2}{a^2} - \frac{(x-h)^2}{b^2} = 1.$$

Do you see the differences between the horizontal and vertical hyperbolas? The x and y switch places (along with the h and k). The a stays on the left, and the b stays on the right. When you write a hyperbola in its standard form, you need to make sure that the positive squared term is always first. The vertices are at $(h, k \pm a)$ and the foci are at $(h, k \pm f)$.

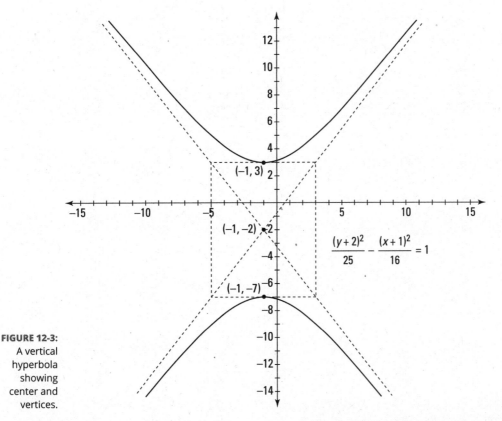

FIGURE 12-3: A vertical hyperbola showing center and vertices.

$$\frac{(y+2)^2}{25} - \frac{(x+1)^2}{16} = 1$$

You can find the asymptotes using the equation $y = \pm \dfrac{a}{b}(x-h)+k$, sort of like the horizontal hyperbolas, just switching the a and b.

Q. Sketch the graph of the hyperbola $16y^2 - 25x^2 + 64y - 50x = 361$.

EXAMPLE **A.** See the following figure. Start by completing the square. Rewrite the equation with the like terms together, and then factor out the coefficients: $16y^2 + 64y - 25x^2 - 50x = 361$ becomes $16(y^2 + 4y) - 25(x^2 + 2x) = 361$. Now, complete the squares (keep the equation balanced, too): $16(y^2 + 4y + 4) - 25(x^2 + 2x + 1) = 361 + 64 - 25$. Factor and simplify: $16(y+2)^2 - 25(x+1)^2 = 400$.

Divide by 400: $\dfrac{(y+2)^2}{25} - \dfrac{(x+1)^2}{16} = 1$. The center of this hyperbola is $(-1, -2)$, where $a = 5$ (and goes up and down) and $b = 4$ (left/right). This is the same hyperbola you saw in the section on Vertical Hyperbolas — you probably didn't recognize it at first.

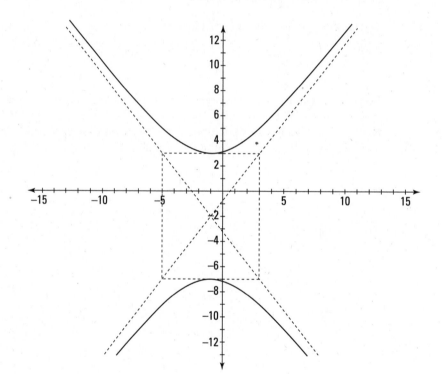

EXAMPLE

Q. You are tossing a chunk of meat into the air to attract a hawk. Your throw follows a hyperbolic curve, reaching a height of 10 feet. The hawk swoops down, but misses at the highest point of the curve by two feet. The distance between where you are standing and where the meat lands is 8 feet. And, amazingly the hawk's trajectory follows the same hyperbolic curve (the meat's mirror image). What is the equation of the hyperbola?

A. $\dfrac{(y-11)^2}{1} - \dfrac{x^2}{2/15} = 1$. Placing the starting and ending points of the tossed meat on the x-axis, you have the hyperbola going through (4, 0) and (−4, 0). The high point is on the y-axis, at (0, 10). Since the hawk missed by 2 feet, it hit a low point of (0, 12).

Looking at the basic form $\dfrac{(y-k)^2}{a^2} - \dfrac{(x-h)^2}{b^2} = 1$, you see that the center must be on the y-axis, making $h = 0$, so you have $\dfrac{(y-k)^2}{a^2} - \dfrac{x^2}{b^2} = 1$. Substituting the points into this equation, you create the system of equations: $\dfrac{k^2}{a^2} - \dfrac{16}{b^2} = 1$, $\dfrac{(10-k)^2}{a^2} = 1$, and $\dfrac{(12-k)^2}{a^2} = 1$.

The solution of the system gives you: $\dfrac{(y-11)^2}{1} - \dfrac{x^2}{2/15} = 1$.

$\dfrac{k^2}{a^2} - \dfrac{16}{b^2} = 1$, $\dfrac{(10-k)^2}{a^2} = 1$, and $\dfrac{(12-k)^2}{a^2} = 1$. The solution of the system gives you:

$\dfrac{(y-11)^2}{1} - \dfrac{x^2}{2/15} = 1$.

---

13 Sketch the graph of the equation $x^2 + 2x - 4y^2 + 32y = 59$.

14 Write the equation of the hyperbola that has its center at $(-3, 5)$, has one vertex at $(-3, 1)$, and passes through the point $\left(1, 5 - 4\sqrt{2}\right)$.

Identifying Conic Sections

Often, you'll be presented with an equation and asked to graph it, but you won't be told what type of conic section it is. It's best to be able to identify what type it is before doing any work. That's easier than it sounds, because there are only four conics, and their equations have distinct differences:

>> **Circles** have x^2 and y^2 with equal coefficients.

>> **Parabolas** have x^2 or y^2, but not both.

>> **Ellipses** have x^2 and y^2 with different (not equal but of the same sign) coefficients.

>> **Hyperbolas** have x^2 and y^2 where exactly one coefficient is negative.

Table 12-1 has all the information you need to know about the four conics in one handy–dandy chart.

Table 12-1   Types of Conic Sections and Their Parts

Type of Conic	Parts of the Conic
Circle	Center (h,k)
$(x-h)^2+(y-k)^2=r^2$	Radius r
Horizontal parabola	Vertex (h,k)
$x=a(y-k)^2+h$	Focus $\left(h+\dfrac{1}{4a},k\right)$
	Directrix $x=h-\dfrac{1}{4a}$
	Axis of symmetry $y=k$
Vertical parabola	Vertex (h,k)
$y=a(x-h)^2+k$	Focus $\left(h,k+\dfrac{1}{4a}\right)$
	Directrix $y=k-\dfrac{1}{4a}$
	Axis of symmetry $x=h$
Horizontal ellipse	Center (h,k)
$\dfrac{(x-h)^2}{a^2}+\dfrac{(y-k)^2}{b^2}=1$	Vertices $(h\pm a,k)$
$f^2=a^2-b^2, a>b$	Co-vertices $(h,k\pm b)$
	Foci $(h\pm f,k)$
Vertical ellipse	Center (h,k)
$\dfrac{(x-h)^2}{b^2}+\dfrac{(y-k)^2}{a^2}=1$	Vertices $(h,k\pm a)$
$f^2=a^2-b^2, a>b$	Co-vertices $b(h\pm a,k)$
	Foci $(h,k\pm f)$

Type of Conic	Parts of the Conic
Horizontal hyperbola	Center (h,k)
$\dfrac{(x-h)^2}{a^2} - \dfrac{(y-k)^2}{b^2} = 1$	Vertices $(h \pm a, k)$
$f^2 = a^2 + b^2$	Foci $(h \pm f, k)$
	Asymptotes $y = \pm \dfrac{b}{a}(x-h) + k$
Vertical hyperbola	Center (h,k)
$\dfrac{(y-k)^2}{a^2} - \dfrac{(x-h)^2}{b^2} = 1$	Vertices $(h, k \pm a)$
$f^2 = a^2 + b^2$	Foci $(h, k \pm f)$
	Asymptotes $y = \pm \dfrac{a}{b}(x-h) + k$

15 Sketch the graph of $3x^2 + 4y^2 - 6x + 16y - 5 = 0$.

16 Sketch the graph of $4x^2 - 8x - 1 = 4y^2 - 4y$.

17 Sketch the graph of $4(x-2) = 2y^2 + 6y$.

18 Sketch the graph of $2y^2 - 4x^2 + 8x - 8 = 0$.

19 Sketch the graph of $4x^2 + 4y^2 - 8y + 16x - 4 = 0$.

20 Sketch the graph of $3x^2 - 4y^2 + 3x - 2y - \dfrac{23}{2} = 0$.

Conic Sections in Parametric Form and Polar Coordinates

So far, you've seen all the conics graphed in *rectangular form* (x, y). However, you can graph a conic section in two other ways:

> » **Parametric form:** This form is for conics that can't be easily written as a function $y = f(x)$. Indeed, unless you have a vertical parabola, a conic will never be expressible as $y = f(x)$. In parametric form, both x and y are written in two different equations as being dependent on one other variable (usually t).

> » **Polar form:** You recognize this from Chapter 11, where every point is expressed as (r, θ).

Both of these forms are dealt with in the following sections.

Parametric form for conic sections

Parametric form defines both x and y in terms of another arbitrary value called the *parameter*. Most often, this is represented by t, as many real-world applications set the definitions based on time. You can find x and y by picking values for t. Why change? In parametric form you can find how far an object has moved over time (the x equation) and the object's height over time (the y equation).

Q. Sketch the curve given by the parametric equations $x = 2t + 1$, $y = t^2 - 3t + 1$, and $1 < t \leq 5$.

EXAMPLE

A. See the following figure. Create a table for t, x, and y. Pick values of t between the interval values given to you, and then figure out what the x and y values are for each t value. Table 12-2 shows these values, and following is the graph of this parametric function.

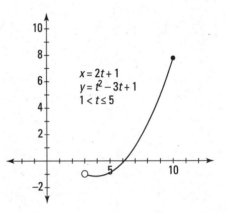

$$x = 2t + 1$$
$$y = t^2 - 3t + 1$$
$$1 < t \leq 5$$

TIP

Even though $t > 1$ in the given interval, you need to start your table off with this value to see what the function would have been. Your graph has an open circle on it at this point to indicate that the value isn't included in the graph or the interval.

Table 12-2   Plug and Chug a Parametric Equation

t value	x value	y value
1	3	−1
2	5	−1
3	7	1
4	9	5
5	11	11

---

21 Sketch the curve defined by the parametric equations $x = t^2 - 1$, $y = 2t$, and $-2 \leq t \leq 3$.

22 Sketch the graph of the parametric equations $x = \dfrac{1}{\sqrt{t-1}}$, $y = \dfrac{1}{t-1}$, and $t > 1$.

Changing from parametric form to rectangular form

Another way to graph a parametric curve is to write it in rectangular form. To do this, you must solve one equation for the parameter and then substitute that value into the other equation. It's easiest if you pick the equation you *can* solve for the parameter (choose the equation that's linear if possible). To show you how it works, I use the example from the last section.

EXAMPLE

Q. Write the parametric equations $x = 2t + 1$, $y = t^2 - 3t + 1$, and $1 < t \leq 5$ in rectangular form.

A. $y = \frac{1}{4}x^2 - 2x + \frac{11}{4}$, x in $(3, 11]$. First, solve the equation that's linear for t: $t = \frac{x-1}{2}$. Then substitute this value into the other equation for t:

$y = \left(\frac{x-1}{2}\right)^2 - 3\left(\frac{x-1}{2}\right) + 1$. Simplify this equation to get $y = \frac{1}{4}x^2 - 2x + \frac{11}{4}$.

---

23 Eliminate the parameter and find an equation in x and y whose graph contains the curve defined by the parametric equations $x = t^2$, $y = 1 - t$, and $t \geq 0$.

24 Eliminate the parameter of the parametric equations $x = t - 5$ and $y = \sqrt{t}$.

Conic sections on the polar coordinate plane

Conic sections on the polar coordinate plane are all based on a special value known as *eccentricity*, or *e*. This value describes what kind of conic section it is, as well as the conic's shape. Knowing what kind of conic section you're dealing with is difficult until you know what the eccentricity is:

» If $e = 0$, the conic is a circle.

» If $0 < e < 1$, the conic is an ellipse.

» If $e = 1$, the conic is a parabola.

» If $e > 1$, the conic is a hyperbola.

When using *e*, all conics are expressed in polar form based on (r, θ), where *r* is the radius and θ is the angle. See Chapter 11 for more information on polar equations.

All conics in polar form are written based on four different equations:

TECHNICAL STUFF

$$r = \frac{ke}{1 - e\cos\theta} \text{ or } r = \frac{ke}{1 - e\sin\theta}$$

$$r = \frac{ke}{1 + e\cos\theta} \text{ or } r = \frac{ke}{1 + e\sin\theta}$$

where *e* is eccentricity and *k* is a constant value. To graph any conic section in polar form, substitute sufficiently many evenly spaced values of θ and graph enough points until you get a picture!

EXAMPLE

Q. Graph the equation $r = \dfrac{2}{4 - \cos\theta}$.

A. See the following figure. First, notice that the equation as shown doesn't fit exactly into any of the equations given in this section. All those denominators begin with 1, and this equation begins with 4! To deal with this, factor out the 4 from the denominator to get

$$r = \frac{2}{4\left(1 - \dfrac{1}{4}\cos\theta\right)}, \text{ which is the same as}$$

$$r = \frac{2 \cdot \dfrac{1}{4}}{1 - \dfrac{1}{4}\cos\theta}.$$

Notice that this makes *e* the same in the numerator and denominator, $\dfrac{1}{4}$, and that *k* is 2. Now that you know *e* is $\dfrac{1}{4}$, that tells you the equation is an ellipse. Plugging in values gives you points $\theta = 0$, $r = \dfrac{2}{3}$; $\theta = \dfrac{\pi}{2}$, $r = \dfrac{1}{2}$; $\theta = \pi$, $r = \dfrac{2}{5}$; $\theta = \dfrac{3\pi}{2}$, $r = \dfrac{1}{2}$. These help you with the shape of the figure.

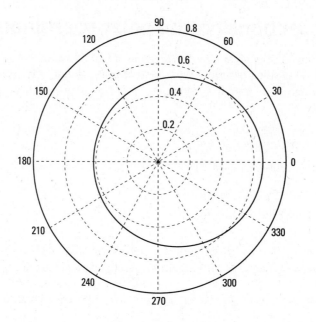

25 Graph the equation of $r = \dfrac{8}{1+\cos\theta}$ and label any vertices.

26 Identify the conic section whose equation is $r = \dfrac{12}{3-4\sin\theta}$ by stating its eccentricity.

Answers to Problems on Conic Sections

This section contains the answers and explanations for the practice problems presented in this chapter.

(1) Find the center and radius of the circle $2x^2 + 2y^2 - 4x = 15$. Then graph the circle. The center is $(1,0)$ and the radius is approximately 2.92.

Rewrite the equation so the x and y variables are together to get $2x^2 - 4x + 2y^2 = 15$. Factor out the coefficient: $2(x^2 - 2x) + 2y^2 = 15$. Complete the square: $2(x^2 - 2x + 1) + 2y^2 = 15 + 2$. Factor and simplify to get $2(x-1)^2 + 2y^2 = 17$. Divide everything by 2 to write the circle in its standard form: $(x-1)^2 + y^2 = 8.5$.

This means the center is $(1, 0)$ and the radius is $\sqrt{8.5}$, or about 2.92.

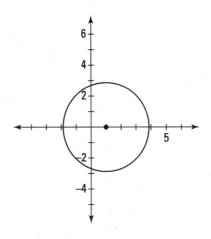

(2) Write the equation of the circle with the center $(-1, 4)$ if the circle passes through the point $(3, 1)$. The answer is $(x+1)^2 + (y-4)^2 = 25$.

If you're given the center and a point on the circle, you can find the distance between the two points using the distance formula, giving you the radius: $d = \sqrt{(-1-3)^2 + (4-1)^2} = \sqrt{16+9} = \sqrt{25} = 5$.

Now that you know both the radius and the center, you can write the equation: $(x+1)^2 + (y-4)^2 = 25$.

(3) What's the vertex of the parabola $y = -x^2 + 4x - 6$? Sketch the graph of this parabola. The answer is $(2, -2)$; see the following graph.

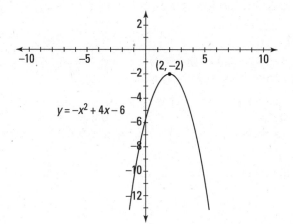

Add the 6 to both sides: $y + 6 = -x^2 + 4x$. Now, factor out the coefficient: $y + 6 = -1\left(x^2 - 4x\right)$. Complete the square and balance the equation: $y + 6 - 4 = -1\left(x^2 - 4x + 4\right)$. Simplify and factor: $y + 2 = -1\left(x - 2\right)^2$. Lastly, subtract 2 from both sides to write the equation in its proper form: $y = -1\left(x - 2\right)^2 - 2$. This means the vertex is located at the point $(2, -2)$.

(4) Find the focus and the directrix of the parabola $y = 4x^2$. The focus is $\left(0, \dfrac{1}{16}\right)$ and the directrix is $y = -\dfrac{1}{16}$.

There's no square to complete, so if it helps you to fill in the missing information with zeros, then rewrite the equation as $y = 4\left(x - 0\right)^2 + 0$. This puts the vertex at the origin $(0,0)$. Because $a = 4$, the focus is $\dfrac{1}{4a}$ units above this point at $\left(0, \dfrac{1}{16}\right)$ and the directrix is the line that runs $\dfrac{1}{4a}$ units below the vertex, perpendicular to the axis of symmetry at $y = -\dfrac{1}{16}$.

(5) Sketch the graph of $x = 2\left(y - 4\right)^2$. See the following graph for the answer.

This equation is written in the proper form, unless you'd like to rewrite it as $x = 2\left(y - 4\right)^2 + 0$ because the h is missing. This is a horizontal parabola with its vertex at $(0, 4)$. It opens to the right, with a steepening times 2.

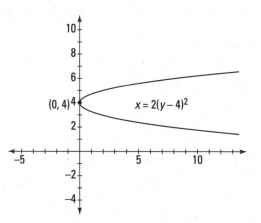

6 Determine whether this parabola opens left or right: $x = -y^2 - 7y + 3$. The answer is the parabola opens to the left.

This problem just asks you whether the parabola opens to the left or right, and you can tell from the leading coefficient of -1 (without doing any work at all) that the parabola opens to the left. If you do actually want to graph this equation, the standard form is $x = -1\left(y + \dfrac{7}{2}\right)^2 + \dfrac{61}{4}$.

7 Sketch the graph of the ellipse $4x^2 + 12y^2 - 8x - 24y = 0$. See the following graph for the answer. You have to complete the square twice. There's no constant to move, so rewrite the equation putting the variables together and factor out the coefficients: $4(x^2 - 2x) + 12(y^2 - 2y) = 0$. Now complete the square and balance the equation: $4(x^2 - 2x + 1) + 12(y^2 - 2y + 1) = 0 + 4 + 12$. Factor and simplify: $4(x - 1)^2 + 12(y - 1)^2 = 16$.

Divide each term by 16: $\dfrac{(x-1)^2}{4} + \dfrac{3(y-1)^2}{4} = 1$. But don't start to graph it yet, because the standard form of an ellipse has a coefficient of 1 in the numerator. Just multiply the numerator and denominator of the second fraction by $\dfrac{1}{3}$ to get $\dfrac{(x-1)^2}{4} + \dfrac{(y-1)^2}{4/3} = 1$.

Now that the equation is written in standard form, you can graph it. The center is $(1, 1)$, the vertices are $(3, 1)$ and $(-1, 1)$, and the co-vertices are $(1, 2.2)$ and $(1, -0.2)$.

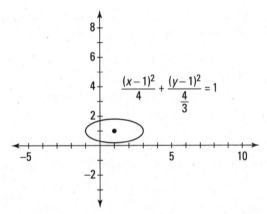

8 Write the equation of the ellipse with vertices at $(-1, 1)$ and $(9, 1)$ and foci at $\left(4 \pm \sqrt{21}, 1\right)$. The answer is $\dfrac{(x-4)^2}{25} + \dfrac{(y-1)^2}{4} = 1$.

Knowing the vertices tells you the center, because it's halfway between them (the midpoint of the segment connecting them — see Chapter 1 for a refresher). This means the center is at $(4, 1)$ and that each vertex is 5 units away from the center, so $a = 5$ and $a^2 = 25$. The foci are $\pm\sqrt{21}$ units away from the center, which tells you that $f = \sqrt{21}$. Now that you know a and f, you can find b^2 using $f^2 = a^2 - b^2$. In this case, $b^2 = 4$. Now you can write the equation: $\dfrac{(x-4)^2}{25} + \dfrac{(y-1)^2}{4} = 1$.

9 Sketch the graph of the ellipse $\dfrac{(x-1)^2}{8} + \dfrac{(y+2)^2}{6} = 1$. See the following graph for the answer.

How convenient! This equation is written in the proper form, so you don't have to complete the square. The center is $(1, -2)$, a is approximately 2.83, and b is approximately 2.45. That gives you the ellipse shown in the graph.

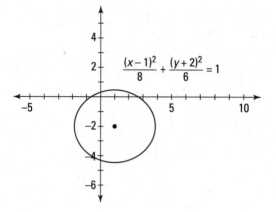

(10) State the ordered pair for the vertices, the co-vertices, and the foci of the ellipse $(x+1)^2 + \dfrac{(y-4)^2}{16} = 1$.

The vertices are $(-1, 8)$ and $(-1, 0)$; the co-vertices are $(-2, 4)$ and $(0, 4)$; and the foci are $\left(-1, 4 \pm \sqrt{15}\right)$. The fact that the x half of the equation isn't written as a fraction is easy to remedy using a denominator of 1: $\dfrac{(x+1)^2}{1} + \dfrac{(y-4)^2}{16} = 1$. This tells you that the center is $(-1, 4)$, $a = 4$, and $b = 1$. The vertices are $(-1, 4 \pm 4) = (-1, 8)$ and $(-1, 0)$. The co-vertices are $(-1 \pm 1, 4) = (-2, 4)$ and $(0, 4)$. Lastly, $f^2 = a^2 - b^2$, so $f = \sqrt{15}$, which gives you the foci at $\left(-1, 4 \pm \sqrt{15}\right)$.

(11) Find the equation of the hyperbola that meets the following criteria: It has its center at $(4, 1)$, one of its vertices is at $(7, 1)$, and one of its asymptotes is $3y = 2x - 5$. The answer is $\dfrac{(x-4)^2}{9} - \dfrac{(y-1)^2}{4} = 1$.

You're given the center $(4, 1)$ and the equation of the asymptote, which you can rewrite in slope-intercept form by dividing by 3 to get $y = \dfrac{2}{3}x - \dfrac{5}{3}$. Because $a = 3$, you see that the vertex is 3 units to the right and this is a horizontal hyperbola. The slope of the asymptote, $\dfrac{2}{3}$, is the value of $\dfrac{b}{a}$. Now that you know the center, a, and b, you can write the equation: $\dfrac{(x-4)^2}{9} - \dfrac{(y-1)^2}{4} = 1$.

(12) Sketch the graph of the equation $(2x - y)(x + 5y) - 9xy = 10$. See the following graph for the answer.

The term $-9xy$ looks a little strange, but it disappears when you FOIL out the binomials and get $2x^2 + 10xy - xy - 5y^2 - 9xy = 10$. Notice that all the xy terms cancel to give you $2x^2 - 5y^2 = 10$. You can divide everything by 10 to write this equation in its form and get $\dfrac{x^2}{5} - \dfrac{y^2}{2} = 1$. The center of this hyperbola is at $(0, 0)$; $a = \sqrt{5}$, or about 2.24; and $b = \sqrt{2}$, or about 1.41.

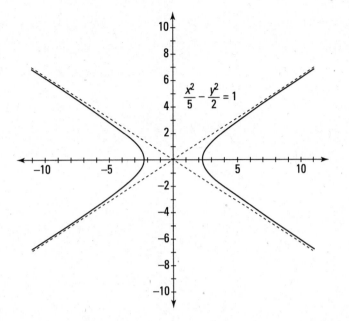

$$\frac{x^2}{5} - \frac{y^2}{2} = 1$$

(13) Sketch the graph of the equation $x^2 + 2x - 4y^2 + 32y = 59$. See the following graph for the answer.

Here's another opportunity to complete the square — twice. Get going by factoring out the coefficients, including the 1 in front of the x^2: $1(x^2 + 2x) - 4(y^2 - 8y) = 59$. Complete the square and keep the equation balanced: $1(x^2 + 2x + 1) - 4(y^2 - 8y + 16) = 59 + 1 - 64$. Factor and simplify: $(x+1)^2 - 4(y-4)^2 = -4$.

Divide each term by -4: $-\frac{(x+1)^2}{4} + \frac{(y-4)^2}{1} = 1$. Did you notice how it suddenly became a vertical hyperbola because the y fraction is positive? Rewriting it to put it in its correct form: $\frac{(y-4)^2}{1} - \frac{(x+1)^2}{4} = 1$. That gives you the following graph.

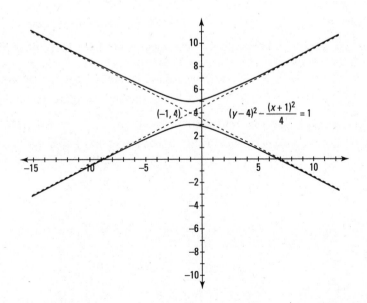

$$(y-4)^2 - \frac{(x+1)^2}{4} = 1$$

14 Write the equation of the hyperbola that has its center at $(-3, 5)$, one vertex at $(-3, 1)$, and passes through the point $\left(1, 5 - 4\sqrt{2}\right)$, which is about $(1, -0.66)$. The answer is $\dfrac{(y-5)^2}{16} - \dfrac{(x+3)^2}{16} = 1$.

Sometimes it helps to draw it out on a sheet of graph paper and mark the center, the vertex, and the point, to see what is going on.

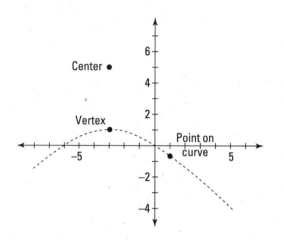

Doing so you see that a vertical hyperbola is what fits, so start with the form for any vertical hyperbola: $\dfrac{(y-k)^2}{a^2} - \dfrac{(x-h)^2}{b^2} = 1$. Based on what you're given, you know that $h = -3$, $k = 5$, and the vertex is 4 units below the center, so $a = 4$. What do you do with the point the question gives you? Remember that all points are (x, y), so if $x = 1$ and $y = 5 - 4\sqrt{2}$, then you can use them to solve for b.

Plug all these values into the equation that you started with and get $\dfrac{\left(5 - 4\sqrt{2} - 5\right)^2}{4^2} - \dfrac{(1+3)^2}{b^2} = 1$, which means you have only one variable to solve for: b^2. This equation simplifies to $2 - \dfrac{16}{b^2} = 1$.

Solving it gets you $b^2 = 16$. This means you can finally write the equation: $\dfrac{(y-5)^2}{16} - \dfrac{(x+3)^2}{16} = 1$.

15 Sketch the graph of $3x^2 + 4y^2 - 6x + 16y - 5 = 0$. See the following graph for the answer.

This is an ellipse because x and y are both squared but have different coefficients of the same sign. Add 5 to both sides and rearrange the terms: $3x^2 - 6x + 4y^2 + 16y = 5$. Factor the coefficients: $3\left(x^2 - 2x\right) + 4\left(y^2 + 4y\right) = 5$. Complete the square and balance the equation: $3\left(x^2 - 2x + 1\right) + 4\left(y^2 + 4y + 4\right) = 5 + 3 + 16$. Factor the perfect square trinomials and simplify: $3(x-1)^2 + 4(y+2)^2 = 24$. Divide everything by 24: $\dfrac{(x-1)^2}{8} + \dfrac{(y+2)^2}{6} = 1$, which turns out to be the same ellipse as in Question 9. This gives you the following graph.

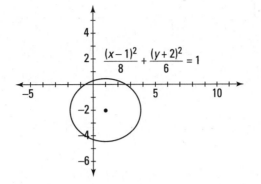

$$\frac{(x-1)^2}{8} + \frac{(y+2)^2}{6} = 1$$

16 Sketch the graph of $4x^2 - 8x - 1 = 4y^2 - 4y$. See the following graph for the answer.

Rewrite the equation first: $4x^2 - 8x - 4y^2 + 4y = 1$. You should recognize that you have a hyperbola on your hands because you have an x^2 and a y^2 where exactly one has a negative coefficient. Factor out the coefficients: $4(x^2 - 2x) - 4(y^2 - y) = 1$. Complete the square and balance the equation: $4(x^2 - 2x + 1) - 4\left(y^2 - y + \frac{1}{4}\right) = 1 + 4 - 1$. Factor the perfect squares and simplify: $4(x-1)^2 - 4\left(y - \frac{1}{2}\right)^2 = 4$. Divide everything by 4: $\frac{(x-1)^2}{1} - \frac{\left(y - \frac{1}{2}\right)^2}{1} = 1$ or $(x-1)^2 - \left(y - \frac{1}{2}\right)^2 = 1$. In this particular hyperbola, the values of a and b are both 1. Knowing this and the center gives you the following graph.

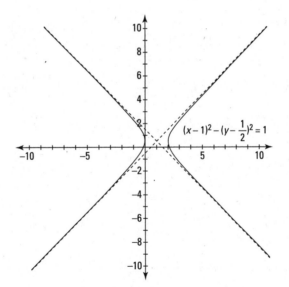

$(x-1)^2 - (y - \frac{1}{2})^2 = 1$

17 Sketch the graph of $4(x-2) = 2y^2 + 6y$. See the following graph for the answer.

Notice right away that the equation doesn't have an x^2, so this is a horizontal parabola. Go ahead and distribute the 4 first: $4x - 8 = 2y^2 + 6y$. Now, factor the coefficient on the y^2 variable: $4x - 8 = 2(y^2 + 3y)$. Completing the square for this one gets you fractions; half of 3 is $\frac{3}{2}$ and that value squared is $\frac{9}{4}$. Add this inside the parentheses, and don't forget to add $2 \cdot \frac{9}{4}$

to the other side to keep the equation balanced: $4x - 8 + \frac{9}{2} = 2\left(y^2 + 3y + \frac{9}{4}\right)$. Simplify and

factor: $4x - \frac{7}{2} = 2\left(y + \frac{3}{2}\right)^2$. Begin to solve for x by adding $\frac{7}{2}$ to each side: $4x = 2\left(y + \frac{3}{2}\right)^2 + \frac{7}{2}$

Now divide each term by 4: $x = \frac{1}{2}\left(y + \frac{3}{2}\right)^2 + \frac{7}{8}$. This gives you a horizontal parabola with a

vertex at $\left(\frac{7}{8}, -\frac{3}{2}\right)$.

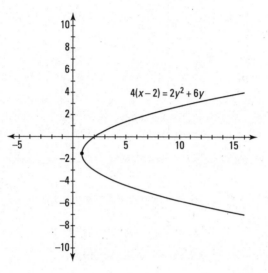

$4(x - 2) = 2y^2 + 6y$

(18) Sketch the graph of $2y^2 - 4x^2 + 8x - 8 = 0$. See the following graph for the answer.

You have a hyperbola to graph this time. Factoring, completing the square and factoring again, you have: $2y^2 - 4\left(x^2 - 2x\right) = 8$ followed by $2y^2 - 4\left(x^2 - 2x + 1\right) = 8 - 4$, which becomes $2y^2 - 4(x-1)^2 = 4$. Dividing by 4, $\frac{y^2}{2} - \frac{(x-1)^2}{1} = 1$. This gives you the hyperbola shown in the graph.

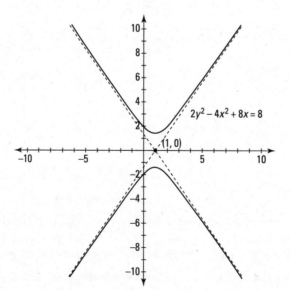

$2y^2 - 4x^2 + 8x = 8$

$(1, 0)$

(19) Sketch the graph of $4x^2 + 4y^2 - 8y + 16x - 4 = 0$. See the following graph for the answer.

You should recognize that this is a circle because of the x^2 and y^2 with equal coefficients on both. Here are the steps to completing the square: $4(x^2 + 4x) + 4(y^2 - 2y) = 4$, then $4(x^2 + 4x + 4) + 4(y^2 - 2y + 1) = 4 + 16 + 4$, which becomes $4(x+2)^2 + 4(y-1)^2 = 24$. Because this is a circle, you need to get coefficients of 1 in front of both sets of parentheses by dividing by 4: $(x+2)^2 + (y-1)^2 = 6$. This circle has its center at $(-2, 1)$ and its radius is $\sqrt{6}$.

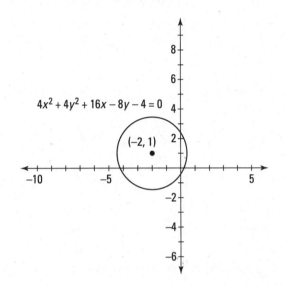

$4x^2 + 4y^2 + 16x - 8y - 4 = 0$

$(-2, 1)$

(20) Sketch the graph of $3x^2 - 4y^2 + 3x - 2y - \dfrac{23}{2} = 0$. See the following graph for the answer.

This is another hyperbola because the coefficient on y^2 is negative while the coefficient on x^2 is positive. Here are the steps used to write the standard form: $3x^2 + 3x - 4y^2 - 2y = \dfrac{23}{2}$, then $3(x^2 + x) - 4\left(y^2 + \dfrac{1}{2}y\right) = \dfrac{23}{2}$, which becomes $3\left(x^2 + x + \dfrac{1}{4}\right) - 4\left(y^2 + \dfrac{1}{2}y + \dfrac{1}{16}\right) = \dfrac{23}{2} + \dfrac{3}{4} - \dfrac{1}{4}$ or

$3\left(x + \dfrac{1}{2}\right)^2 - 4\left(y + \dfrac{1}{4}\right)^2 = 12$. Dividing each term by 12, $\dfrac{\left(x + \dfrac{1}{2}\right)^2}{4} - \dfrac{\left(y + \dfrac{1}{4}\right)^2}{3} = 1$ gives you the hyperbola shown in the graph.

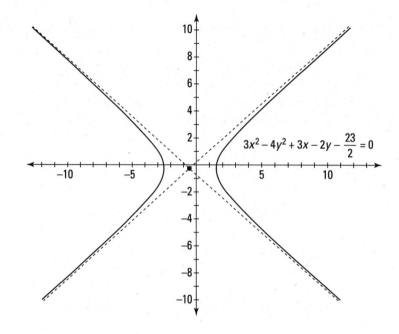

$$3x^2 - 4y^2 + 3x - 2y - \frac{23}{2} = 0$$

21 Sketch the graph of the parametric equations $x = t^2 - 1$, $y = 2t$, and $-2 \le t \le 3$. See the following graph for the answer.

Set up a table of t, x, and y where you pick the t and find x and y. Be sure to stay within the interval defined by the problem. Here's your chart:

t	x	y
-2	3	-4
-1	0	-2
0	-1	0
1	0	2
2	3	4
3	8	6

These (x, y) points give you points on the graph of the parabola $x = \left(\dfrac{y}{2}\right)^2 - 1$ or $x = \dfrac{1}{4}y^2 - 1$. You can solve for this equation by substituting $t = \dfrac{y}{2}$ into $x = t^2 - 1$.

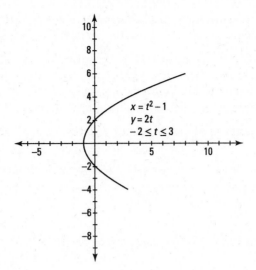

$x = t^2 - 1$
$y = 2t$
$-2 \le t \le 3$

22 Sketch the graph of the parametric equations $x = \dfrac{1}{\sqrt{t-1}}$, $y = \dfrac{1}{t-1}$ and $t > 1$. See the following graph for the answer.

Another chart comes in handy here:

t	x	y
1	undefined	undefined
1.25	2	4
2	1	1
3	$\dfrac{1}{\sqrt{2}}$	$\dfrac{1}{2}$
4	$\dfrac{1}{\sqrt{3}}$	$\dfrac{1}{3}$
5	$\dfrac{1}{2}$	$\dfrac{1}{4}$

These points are on the graph of $y = x^2$. You can find this by substituting $t = 1 + \dfrac{1}{y}$ into $x = \dfrac{1}{\sqrt{t-1}}$ and then solving for y.

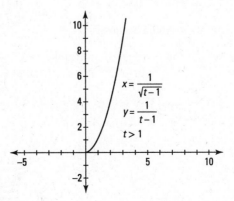

$x = \dfrac{1}{\sqrt{t-1}}$
$y = \dfrac{1}{t-1}$
$t > 1$

(23) Eliminate the parameter and find an equation in x and y whose graph contains the curve of the parametric equations $x = t^2$, $y = 1 - t$, and $t \geq 0$. The answer is $y = 1 - \sqrt{x}$.

Solve for t in the first equation by taking the square root of both sides: Since $t \geq 0$, $\sqrt{x} = t$. Substitute this value into the other equation: $y = 1 - \sqrt{x}$. You can then continue to put the equation into the form of a conic. To do this, subtract 1 from both sides: $y - 1 = -\sqrt{x}$. Then square both sides to get $(y-1)^2 = x$. You have a parabola.

(24) Eliminate the parameter of the parametric equations $x = t - 5$ and $y = \sqrt{t}$. The answer is $y = \sqrt{x+5}$.

The first equation is easy to solve for t by adding 5 to both sides: $x + 5 = t$. Substitute this value into the other equation and get $y = \sqrt{x+5}$. If you'd rather write this equation without the square root, square both sides to get $y^2 = x + 5$. This, too, looks like a portion of a parabola.

(25) Graph the equation of $r = \dfrac{8}{1+\cos\theta}$ and label any vertices. See the following graph for the answer.

Notice that the given equation is the same thing as $r = \dfrac{8 \cdot 1}{1 + 1 \cdot \cos\theta}$, which tells you that $k = 8$ and $e = 1$. This makes it a parabola.

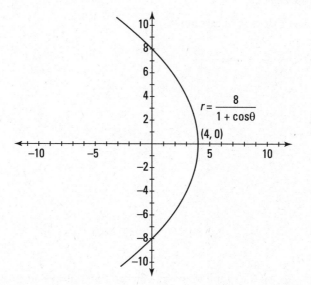

(26) Identify the conic section whose equation is $r = \dfrac{12}{3 - 4\sin\theta}$ by stating its eccentricity. The answer is this conic is a hyperbola because $e = \dfrac{4}{3}$.

You have to factor the 3 out of the denominator first: $r = \dfrac{12}{3\left(1 - \dfrac{4}{3}\sin\theta\right)}$. This is the same thing as $r = \dfrac{\dfrac{12}{3}}{1 - \dfrac{4}{3}\sin\theta} = \dfrac{3 \cdot \dfrac{4}{3}}{1 - \dfrac{4}{3}\sin\theta}$, and that tells you that $k = 3$ and $e = \dfrac{4}{3}$, which is why this one is a hyperbola.

Chapter **13**

Finding Solutions for Systems of Equations

No, a system of equations is *not* a way to organize, arrange, or classify equations. A *system of equations* is a collection of equations involving the same set of variables. The point is to find all solutions, if they exist, that work in all the equations. Solving one equation for one variable is often your first and wisest option.

It goes without saying that the bigger the system of equations becomes, the more challenging it may be to solve. Solving a system involves several options, and sometimes it may be easier to solve certain systems certain ways.

Of course, you can choose to always solve all systems using one technique, but another technique may require fewer steps and assure you of more accuracy.

A Quick-and-Dirty Technique Overview

Here's a handy guide to all the techniques covered in this chapter and when it's best to use them:

>> If a system has two or three variables, you may be able to use *substitution* or *elimination* to solve.

>> If a linear system has four or more variables, you can use *matrices,* in which case you have the following choices:

- The Gaussian method
- Inverse matrices
- Cramer's Rule

All these techniques are discussed in detail in this chapter. And, in addition, you'll also find *systems of inequalities* and how to solve them by graphing.

Solving Two Linear Equations with Two Variables

When you're presented with a system of linear equations with two variables, the best methods to solve them are known as *substitution* and *elimination*. As mentioned earlier in this chapter, you can use either method to solve almost any system of this type, and you're shown both methods, because each one has its unique advantages. In keeping with this spirit, here are both methods, and notes on *when* to use each one as well.

REMEMBER

Just remember that with each system of equations with two variables, you need to find the value of *both* variables, usually x and y. Don't stop until you've solved for both, or you haven't really solved the system.

REMEMBER

Also remember that sometimes, systems of equations don't have a solution. For example, sometimes the two lines representing the system are parallel to each other — without an intersection! This means that there is *no solution.* The fact that there may be no solution may pop up from time to time using these other methods as well.

So how do you recognize a system of equations with no solution without using a graph? That's easy — when doing your work, you end up with an equation that just doesn't make sense. It may say $2 = 7$ or $-1 = 10$; that tells you right away that it has no solution. It's also possible to be given the same line (in disguise) twice. If you were to graph that system, you'd end up with one line on top of another.

These two lines share infinitely many points, so you say that the system has infinitely many solutions. These equations boil down at some point to an *identity* — the left and right sides of the equation are exactly the same (such as $2 = 2, 10x = 10x$, or $4y - 3 = 4y - 3$), and these, too, are easy to recognize.

The substitution method

In the *substitution method,* you solve one equation for one variable and then substitute the expression for that variable in the other equation. If one of the two equations you're given has already been solved for one variable, then take advantage of the situation. You know you've got a winner for the substitution method. Of course, if one equation can be easily solved for one variable (one variable has a coefficient of 1), you also know that substitution is a good bet in this case, also.

Q. Solve the system of equations:

$$x = 4y - 1$$
$$2x + 5y = 11$$

A. $x = 3, y = 1$. Notice how the first equation says "$x = ...$"? This just begs you to use substitution. You can take this first expression and substitute it for the x in the other equation. This gets you $2(4y-1)+5y = 11$. The substitution method makes your job easier because you end up with one equation in one variable — and this one is easy to solve! When you do, you get $y = 1$. Now that you know half of your answer (y in this case), you can substitute that value into the original equation to get the other half (x).

Save yourself some time and steps by substituting the first answer you get into the equation that has already been solved for a variable. For this example, because you know that $x = 4y - 1$ and you've figured out that $y = 1$, it takes very few steps to figure out that $x = 4(1) - 1 = 4 - 1 = 3$. Of course, you should always check your answer in the other equation — substituting the values you found in for x and y.

 Use substitution to solve the system
$$\begin{cases} x + y = 6 \\ y = 13 - 2x \end{cases}$$

2 The sum of two numbers is 14 and their difference is 2. Find the numbers.

The elimination method

TECHNICAL STUFF

Elimination is the method of choice when both of the linear equations given to you are written in *standard form:*

$$Ax + By = C$$
$$Dx + Ey = F$$

where A, B, C, D, E, and F are all real numbers. It's called *standard form* to distinguish it from the slope–intercept form or other preferred formats.

In the two equations, if the coefficients of either the x or y terms are opposites of each other, you should choose elimination. In the *elimination method,* you add the two equations together so that one of the variables disappears (is eliminated). Sometimes, however, you must multiply one or both equations by a constant in order for the coefficients to have opposite signs. This way, when you add the two equations together, one of the variables will be eliminated.

EXAMPLE

Q. Solve the system $\begin{cases} 2x - y = 6 \\ 3x + y = 4 \end{cases}$

A. $x = 2, y = -2$. Notice that you could solve this system using substitution, because in the second equation, you can solve for y by subtracting $3x$ from both sides. But, also, the y terms are exact opposites of each other. If you add the two equations, you get $5x = 10$. This means that you can divide both sides by 5 to easily solve for x and get $x = 2$. Substituting this back into either equation, you find that $y = -2$.

Q. Solve the system $\begin{cases} 2x - 3y = 5 \\ 4x + 5y = -1 \end{cases}$

A. $x = 1, y = -1$. Solving this system by substitution is a possible choice, but it would eventually mean dividing through by one of the coefficients and creating fractions. Instead, you can avoid the fractions by using the elimination method. The fact that both equations are written in standard form is another vote in favor of the elimination method. Notice that the y terms have opposite signs, so you can eliminate them (you can eliminate any variable you choose, but it's all about taking the fewest steps). It's a little like finding the least common multiple of both coefficients, in this case, the 3 and the 5. The smallest number that both of those go into is 15, so you have to multiply the top equation by 5 and the bottom equation by 3. This gives you $\begin{cases} 10x - 15y = 25 \\ 12x + 15y = -3 \end{cases}$. Adding these two equations together gives you $22x = 22$, which gives you the solution $x = 1$. You then substitute this value back into one of the two original equations to solve for y. In this example, $y = -1$.

3 Solve the system $\begin{cases} \dfrac{x}{2} - \dfrac{y}{3} = -3 \\ \dfrac{2x}{3} + \dfrac{y}{2} = -2 \end{cases}$

4 Solve the system $\begin{cases} 3x - 2y = 4 \\ 6x - 4y = 8 \end{cases}$

Not-So-Straight: Solving Nonlinear Systems

The substitution and elimination methods are common tools for systems of equations that include *nonlinear equations*. Yes, now at least one of your two given equations can be a quadratic equation (it may also be a rational function or some other type). The method you choose to use for these types of systems depends on the types of equations that you're given. The following sections are broken into those types with recommendations on how to best solve each one.

One equation that's linear and one that isn't

TIP

When one equation is linear and the other equation isn't linear, it's usually best to use the substitution method. That's because the linear equation can be easily solved for one variable. You can then substitute this value into the other equation to solve. Quite often, this means solving a quadratic equation at some point. If you need to brush up on those techniques, see Chapter 4.

EXAMPLE

Q. Solve the system of equations
$\begin{cases} x^2 + y = 0 \\ 2x - y = 3 \end{cases}$

A. $x = -3$, $y = -9$ and $x = 1$, $y = -1$. This system involves a parabola and a line, so you can expect as many as two points of intersection. It's usually easier to solve for a variable in the linear equation first. The second given equation is the linear one, and it's easier to solve for y (no coefficients to divide). Doing so gets you $y = 2x - 3$. After you substitute this expression into the first equation for y, you get $x^2 + 2x - 3 = 0$. This quadratic polynomial factors to $(x+3)(x-1) = 0$. Then, using the zero product property (for more information, see Chapter 4), you get two solutions: $x = -3$ or $x = 1$. Two solutions for x means twice the substitution and twice the y answers. If $x = -3$, then $y = -9$, and if $x = 1$, $y = -1$.

5 Solve $\begin{cases} x^2 - y = 1 \\ x + y = 5 \end{cases}$

6 Solve $\begin{cases} x + y = 9 \\ xy = 20 \end{cases}$

Two nonlinear equations

In this section, you see what happens when both of the given equations are nonlinear. When elimination can be used, it's really the preferred method for solving the system, but substitution is a great fall back.

EXAMPLE

Q. Solve the system $\begin{cases} x^2 + y^2 = 25 \\ x^2 - y = 5 \end{cases}$

A. $(0, -5)$, $(3, 4)$, and $(-3, 4)$. This involves a circle and parabola, so you can have as many as four points of intersection. Notice that the x^2 terms in both equations have the same coefficient. If you multiply the second equation by -1 and then add the two equations together, you get $y^2 + y = 20$. Rewrite

the quadratic equation, $y^2 + y - 20 = 0$, that you can factor: $(y + 5)(y - 4) = 0$. Solve and get $y = -5$ and $y = 4$. Substituting $y = -5$ into the second equation gets you $x^2 = 0$, which means that $x = 0$. Substituting $y = 4$ into the same equation gets you $x^2 - 4 = 5$ or $x^2 = 9$, which gives you $x = \pm 3$. Both of these solutions work.

7 Solve $\begin{cases} x^2 + y^2 = 1 \\ x + y^2 = -5 \end{cases}$

8 Solve $\begin{cases} 27x^2 - 16y^2 = -400 \\ 4y^2 - 9x^2 = 36 \end{cases}$

Systems of rational equations

WARNING

Sometimes you'll see two equations that consist of rational equations. As you see in Chapter 3, sometimes rational functions have undefined values. Keep this in mind with the final solutions you find when doing these problems — they may not really work! *Always* check the solutions to these types of equations because you never know which ones are actually solutions and which ones aren't until you double-check.

EXAMPLE

Q. Solve the system $\begin{cases} \dfrac{3}{x} + \dfrac{2}{y} = 11 \\ \dfrac{1}{x} - \dfrac{1}{y} = -11 \end{cases}$

A. $x = -\dfrac{5}{11}, y = \dfrac{5}{44}$. With both x and y in the denominator, neither of them can be equal to 0, so that's the only thing you have to watch out for in your solution. The two fractions with denominators y can be eliminated if you multiply the second equation through by 2 and add the equations together. The system is now written with $\dfrac{3}{x} + \dfrac{2}{y} = 11$ and $\dfrac{2}{x} - \dfrac{2}{y} = -22$ and the sum is $\dfrac{5}{x} = -11$. Solving for x, you get $x = -\dfrac{5}{11}$. Substituting that into the first equation, $\dfrac{3}{-5/11} + \dfrac{2}{y} = 11$ becomes $-\dfrac{33}{5} + \dfrac{2}{y} = 11$ or $\dfrac{2}{y} = 11 + \dfrac{33}{5}$. Solving for y, you get $y = \dfrac{5}{44}$.

---

9 Solve $\begin{cases} \dfrac{14}{x+3} + \dfrac{7}{4-y} = 9 \\ \dfrac{21}{x+3} - \dfrac{3}{4-y} = 0 \end{cases}$

10 Solve $\begin{cases} \dfrac{12}{x+1} - \dfrac{12}{y-1} = 8 \\ \dfrac{6}{x+1} + \dfrac{6}{y-1} = -2 \end{cases}$

Systems of More Than Two Equations

Systems of equations may have more than three equations and/or variables. You can sometimes draw three-dimensional figures of the three-equation situations, but larger systems don't cooperate on the picture front.

TIP

Most of the time when you're given systems larger than 2×2, you want to use elimination. You take two equations at a time and eliminate one variable. Then, you take another two equations and eliminate the same variable. If you start with a 3×3 system, this knocks you down to a 2×2 system, which you then can solve.

EXAMPLE

Q. Solve the system of equations

$$\begin{cases} x - y + z = 6 \\ x - z = -2 \\ x + y = -1 \end{cases}$$

A. $x = 1, y = -2, z = 3$. Notice that the second two equations have only two variables. This helps you decide which variable to eliminate in the three equations. If you prefer working with just x and y, then use the first and second equations and eliminate the z. Adding those two equations together, you get $2x - y = 4$. Now you can create the system of two equations using that result and the original third equation. $\begin{cases} 2x - y = 4 \\ x + y = -1 \end{cases}$. Adding these two equations together, you get $3x = 3$ or $x = 1$. Put that value into the original third equation, and you find that $y = -2$. And those two values in the original first equation give you that $z = 3$.

Q. Solve the system $\begin{cases} x + y + z + w = -1 \\ 2x + y + z = 0 \\ 2y + z - w = -6 \\ x - z + 2w = 7 \end{cases}$

A. $x = 2, y = -1, z = -3, w = 1$. You want to eliminate one of the variables in all four equations. Since the w is missing in the second equation, a good move would be to choose that variable. Adding the first and third equations together, you get $x + 3y + 2z = -7$. Then, multiplying the third equation by 2 and adding the result to the fourth equation you have $x + 4y - z = -5$. Your newly created system of three equations is $\begin{cases} 2x + y + z = 0 \\ x + 3y + 2z = -7 \\ x + 4y + z = -5 \end{cases}$. Now choose z to eliminate. Multiplying the first equation in this reduced system by -2 and adding it to the second equation, you get $-3x + y = -7$, and multiplying the third equation by -1 and adding it to the first, you get $x - 3y = 5$. Finally, multiplying the newest first equation by 3 and adding it to the newest second equation, you get $-8x = -16$ or $x = 2$. Back solving, you get then that $y = -1$, $z = -3$, and $w = 1$.

(11) Solve the system $\begin{cases} 3x - 2y = 17 \\ x - 2z = 1 \\ 3y + 2z = 1 \end{cases}$

(12) Solve $\begin{cases} 2x - y + z = 1 \\ x + y - z = 2 \\ -x - y + z = 2 \end{cases}$

(13) Solve $\begin{cases} 2x + 3y + 4z = 37 \\ 4x - 3y + 2z = 17 \\ x + 2y - 3z = -5 \\ 3x - 2y + z = 11 \end{cases}$

(14) Solve $\begin{cases} 3a + b + c + d = 0 \\ 4a + 5b + 2c = 15 \\ 4a + 2b + 5d = -10 \\ -5a + 3b - d = 8 \end{cases}$

Graphing Systems of Inequalities

A *system of inequalities* can contain two or more inequalities and two or more variables. The first inequalities you're introduced to are usually linear. In later studies, you continue with those types of problems but then move up to nonlinear systems of inequalities. For a review of how to graph one inequality, see Chapter 1. Graphing quadratics is found in Chapter 3, and conics in Chapter 12. Take a look at these different graphs if you need to review before proceeding.

REMEMBER

One way to solve a system of inequalities is to graph it. You end up with (hopefully) two overlapping shaded regions — the overlap is the solution. Every single point in the overlap is a solution to the system. What happens if there's no overlap? Well, there's no solution!

WARNING

When you multiply or divide an inequality by a negative number, the inequality sign changes: < becomes >, ≤ becomes ≥, and vice versa. This is a pretty important fact to remember because it affects your shading in the end.

EXAMPLE

Q. Sketch the graph of the system of inequalities:

$$\begin{cases} 3x + y \le 5 \\ x + 2y \le 4 \end{cases}$$

A. See the following figure. Because both of these inequalities are linear, you should put them in slope-intercept form to graph them. The system is then written $\begin{cases} y \le -3x + 5 \\ y \le -\dfrac{1}{2}x + 2 \end{cases}$.

Graphing both on the same coordinate plane gives you the figure. You shade below and to the left of the two lines to include the point $(0, 0)$.

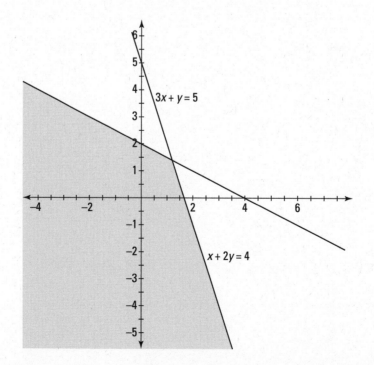

Q. Sketch the graph of the system of inequalities

$$\begin{cases} x^2 + y^2 < 16 \\ x - 2y > -4 \end{cases}$$

EXAMPLE

A. See the following figure. This time, the top inequality represents the interior of a circle, and the bottom inequality represents a half-plane (bounded by a line). The circle is in the proper form to graph, so you don't have to change it. The bottom inequality should be put in slope-intercept form is $y < \dfrac{1}{2}x + 2$. For these types of problems, I recommend that you pick test points to see where to shade. For example, the origin $(0, 0)$ is a great point to try in the original inequalities to see whether it works. Using the test point $(0, 0)$, you find that you shade inside the circle and below the line.

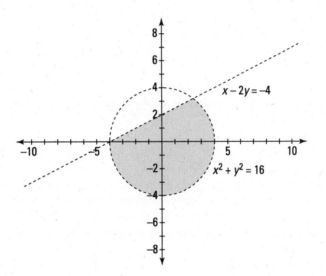

15 Sketch the graph of $\begin{cases} 2x + y \geq 9 \\ 2x - y \geq 1 \\ x \leq 7 \end{cases}$

16 Sketch the graph of $\begin{cases} x^2 + y^2 \geq 9 \\ x^2 + (y-3)^2 \geq 9 \end{cases}$

17 Sketch the graph of $\begin{cases} x^2 - y > 2 \\ x - y < 4 \end{cases}$

18 Sketch the graph of $\begin{cases} y \geq 0 \\ x + y < 4 \\ y \leq \sqrt{x-1} \end{cases}$

Breaking Down Decomposing Partial Fractions

Some people are partial to partial fractions! (Sorry. Couldn't resist.) The process known as decomposing into partial fractions involves taking one fraction and breaking it down into the sum or difference of two (or more) other fractions. The method is often called just *partial fractions*. This process requires some factoring, so if you need a review, turn to Chapter 4 and read up on how to do it. To perform the process, just follow these general steps as long as the degree of the numerator is less than that of the denominator (otherwise, do long division first):

TECHNICAL STUFF

1. **Factor the denominator.**

2. **Write separate fractions, one for each factor of the denominator based on these rules:**

   • If the factor is linear, it has some constant in the numerator.

   • If the factor is quadratic and irreducible (doesn't have real roots), it has a linear expression in the numerator.

Note: If any factor has a power on it, you create one fraction with that power, and then additional fractions with the factor to all the lesser powers down to a power of 1. This is probably best shown with an example. Suppose that you're able to factor the denominator of a fraction into $\dfrac{\text{numerator}}{(x-4)(x+1)^3(2x-1)(3x^2-4x+2)^2}$. You create the sum of seven different fractions:

$$\frac{A}{(x-4)} + \frac{B}{(x+1)^3} + \frac{C}{(x+1)^2} + \frac{D}{(x+1)^1} + \frac{E}{(2x-1)} + \frac{Fx+G}{(3x^2-4x+2)^2} + \frac{Hx+J}{(3x^2-4x+2)^1}$$

The first denominator factor is linear, so its numerator is a constant. The second denominator factor is linear with multiplicity 3, so you need to create three different constant numerators:

one for the third degree, the second for the second degree, and the third for the first degree. The third factor is also linear, so it gets one fraction with a constant on the top. Lastly, the final term is quadratic with multiplicity 2, so it gets two linear terms in the numerator: one for the second degree and the second for the first degree.

EXAMPLE

Q. Find the partial fraction decomposition of $\dfrac{7x+5}{x^2+x-2}$.

A. $\dfrac{4}{x-1}+\dfrac{3}{x+2}$. Begin by factoring the denominator of the given fraction: $\dfrac{7x+5}{(x-1)(x+2)}$. Because both factors in the denominator are linear, you write the fraction as the sum of two different fractions with constant numerators: $\dfrac{7x+5}{(x-1)(x+2)}=\dfrac{A}{x-1}+\dfrac{B}{x+2}$. Now, multiply each fraction in the equation by the factored denominator, which results in some big-time canceling:

$$\dfrac{7x+5}{\cancel{(x-1)}\cancel{(x+2)}}\cdot\cancel{(x-1)}\cancel{(x+2)}=\dfrac{A}{\cancel{x-1}}\cdot\cancel{(x-1)}(x+2)+\dfrac{B}{\cancel{x+2}}\cdot(x-1)\cancel{(x+2)}.$$ This simplifies

to $7x+5=A(x+2)+B(x-1)$. When you multiply this out, you get the equation $7x+5=Ax+2A+Bx-B$. Now, gathering like terms and factoring: $7x+5=(A+B)x+(2A-B)$. Notice how both sides match up, with a term multiplying x and a constant. You can write the equations: $7x=(A+B)x$ or $7=A+B$ and $5=2A-B$. This gives you a system of equations to solve: $\begin{cases}7=A+B\\5=2A-B\end{cases}$. Solving the system by adding the two equations together, you first get $12=3A$ and then $4=A$. Substituting this into the first equation, you have $B=3$. Take these values back to $\dfrac{7x+5}{(x-1)(x+2)}=\dfrac{A}{x-1}+\dfrac{B}{x+2}$, and you have that $\dfrac{7x+5}{(x-1)(x+2)}=\dfrac{4}{x-1}+\dfrac{3}{x+2}$.

---

19 Find the constants A and B such that
$$\dfrac{x-38}{x^2+x-12}=\dfrac{A}{x+4}+\dfrac{B}{x-3}.$$

20 Find the initial form of the partial fraction decomposition for $\dfrac{5x-4}{(x-1)^2}$ (but don't solve for the constants).

21 Find the partial fraction decomposition for

$$\frac{2x^2 - 21x + 18}{(x-1)(x^2 - 4x + 4)}$$

22 Find the partial fraction decomposition for

$$\frac{11x^2 - 7x + 14}{2x^3 - 4x^2 + 3x - 6}$$

Working with a Matrix

A *matrix* is a collection of numbers or other objects arranged in rows and columns. Each number inside the matrix is called an *entry* or *element*. A matrix comes in handy when you have a bunch of data that you need to organize and keep track of. Capital letters are used to name matrices so you can keep track of which matrix you're referring to at the time. And one way a matrix is classified is by its *dimensions*, or how big it is. This is also sometimes known as the *order* of the matrix and is always the number of rows by the number of columns. For example, if matrix M is 4×3, it has four rows and three columns. After the data is organized in this fashion, you can add, subtract, and even multiply matrices. There's also an operation known as *scalar multiplication*, which means you multiply the entire matrix by a constant.

REMEMBER To add or subtract matrices, you have to operate on their corresponding elements. In other words, you add or subtract the element in the first row/first column in one matrix to or from the exact same element in another matrix. The two matrices must have the same dimensions; otherwise, an element in one matrix won't have a corresponding element in the other.

The following example shows two matrices and what their sum and differences are:

$$A = \begin{bmatrix} -5 & 1 & -3 \\ 6 & 0 & 2 \\ 2 & 6 & 1 \end{bmatrix} \qquad B = \begin{bmatrix} 2 & 4 & 5 \\ -8 & 10 & 3 \\ -2 & -3 & -9 \end{bmatrix}$$

$$A + B = \begin{bmatrix} -3 & 5 & 2 \\ -2 & 10 & 5 \\ 0 & 3 & -8 \end{bmatrix} \qquad A - B = \begin{bmatrix} -7 & -3 & -8 \\ 14 & -10 & -1 \\ 4 & 9 & 10 \end{bmatrix}$$

Next, you see scalar multiplication 3A:

$$3A = 3\begin{bmatrix} -5 & 1 & -3 \\ 6 & 0 & 2 \\ 2 & 6 & 1 \end{bmatrix} = \begin{bmatrix} -15 & 9 & -9 \\ 18 & 0 & 6 \\ 6 & 18 & 3 \end{bmatrix}$$

Multiplying matrices is another issue. First of all, to multiply two matrices AB (the matrices are written right next to each other, with no symbol in between), the number of columns in matrix A *must* match the number of rows in matrix B. If matrix A is $m \times n$ and matrix B is $n \times p$, the product AB has dimensions $m \times p$. And also, when it comes to matrix multiplication, AB doesn't equal BA; in fact, just because AB exists doesn't even mean that BA does as well.

For the two example questions in this section, let

$$C = \begin{bmatrix} -5 & -1 & 3 & 6 \\ 0 & 2 & -2 & 6 \end{bmatrix} \quad D = \begin{bmatrix} 2 & 4 & 5 & -8 \\ 10 & 3 & -2 & -3 \end{bmatrix} \quad E = \begin{bmatrix} -1 & 2 & -1 \\ 4 & 4 & 0 \\ 2 & 3 & 1 \\ -5 & 2 & -1 \end{bmatrix}$$

Q. Find $3C - 2D =$

A. $3C - 2D = \begin{bmatrix} -19 & -11 & -1 & 34 \\ -20 & 0 & -2 & 24 \end{bmatrix}$

To find $3C - 2D = 3\begin{bmatrix} -5 & -1 & 3 & 6 \\ 0 & 2 & -2 & 6 \end{bmatrix} - 2\begin{bmatrix} 2 & 4 & 5 & -8 \\ 10 & 3 & -2 & -3 \end{bmatrix}$, follow the order of operations

and multiply by the scalars: $= \begin{bmatrix} -15 & -3 & 9 & 18 \\ 0 & 6 & -6 & 18 \end{bmatrix} - \begin{bmatrix} 4 & 8 & 10 & -16 \\ 20 & 6 & -4 & -6 \end{bmatrix}$. Then subtract the

two matrices, watching the negative signs: $= \begin{bmatrix} -19 & -11 & -1 & 34 \\ -20 & 0 & -2 & 24 \end{bmatrix}$.

Q. Find CE.

A. $CE = \begin{bmatrix} -23 & 7 & 2 \\ -26 & 14 & -8 \end{bmatrix}$

$$CE = \begin{bmatrix} -5 & -1 & 3 & 6 \\ 0 & 2 & -2 & 6 \end{bmatrix} \cdot \begin{bmatrix} -1 & 2 & -1 \\ 4 & 4 & 0 \\ 2 & 3 & 1 \\ -5 & 2 & -1 \end{bmatrix}$$

You need to multiply each element of each row of the left matrix by each element of each column of the right matrix and add up the products.

The sum of the first row times the first column: $-5(-1) - 1(4) + 3(2) + 6(-5) = -23$. This is the first row, first column answer.

The sum of the first row times the second column: $-5(2) - 1(4) + 3(3) + 6(2) = 7$. This is the first row, second column answer.

The sum of the first row times the third column: $-5(-1) - 1(0) + 3(1) + 6(-1) = 2$. This is the first row, third column answer.

The sum of the second row times the first column: $0(-1)+2(0)-2(1)+6(-5)=-26$. This is the second row, first column answer.

The sum of the second row times the second column: $0(2)+2(4)-2(3)+6(2)=14$. This is the second row, second column answer.

The sum of the second row times the third column: $0(-1)+2(0)-2(1)+6(-1)=-8$. This is the second row, third column answer.

Putting these all into a matrix gives you the answer $CE = \begin{bmatrix} -23 & 7 & 2 \\ -26 & 14 & -8 \end{bmatrix}$

---

23 Find 4D.

24 Find $4D + 5C$.

25 Find $3C - E$.

26 Find DE.

Getting It in the Right Form: Simplifying Matrices

You can write a linear system of equations in *matrix form*. To do so, follow these steps:

1. **Write all the coefficients in one matrix, called the *coefficient matrix*. Each equation gets its own row in the matrix, and each variable gets its own column, written in the same order as the equations.**

2. **Write another matrix (column matrix) with all the variables in it, called the *variable matrix*, in order from top to bottom. Show this being multiplied by the coefficient matrix.**

3. **Set this product equal to a column matrix with the answers in it, sometimes called the *answer matrix*.**

Row echelon form is a special designation for a matrix; when using this form, you only work with the coefficient matrix. Across any row, the first number element (besides 0) that you run into is called the *leading coefficient*. For a coefficient matrix to be in row echelon form,

TECHNICAL STUFF

>> Any row with all 0s in it must be the bottom row.

>> The leading coefficient in any row must be to the right of the leading coefficient in the row above it.

Reduced row echelon form takes row echelon form and makes all the leading coefficients the number 1. Also, each element above or below a leading coefficient must be 0. The following matrix A is in reduced row echelon form, while the following matrix B is in row echelon but not in *reduced* row echelon form.

$$A = \begin{bmatrix} 0 & 1 & 0 & 0 \\ 0 & 0 & 1 & 0 \\ 0 & 0 & 0 & 1 \end{bmatrix} \qquad B = \begin{bmatrix} 0 & 1 & 2 & 7 \\ 0 & 0 & 1 & 0 \\ 0 & 0 & 0 & 1 \end{bmatrix}$$

Finally, *augmented form* takes the coefficient matrix and tacks on an extra column — a column with the answers in it so that you can look at the entire system in one convenient package.

These ways of writing systems of equations in matrices come in handy when dealing with systems that are 4×4 or larger. Your goal is to get the matrix into row echelon form using *elementary row operations*. Here are three row operations you can perform:

TECHNICAL STUFF

>> Multiply each element of a row by a constant.

>> Interchange any two rows.

>> Add two rows together.

You'll find a full explanation of these operations in *Pre-Calculus For Dummies*, using the same notation to represent these elementary row operations. The notation $4r_2 \rightarrow r_2$ indicates that 4 times row two is replacing the original row two. The notation $r_1 \leftrightarrow r_3$ says that you swap row one and row three. Then $r_3 + r_1 \rightarrow r_1$ adds row three to row one and replaces row one with that sum. And $4r_2 + r_1 \rightarrow r_1$ first multiplies row two by 4 and then adds that to row one and replaces row one with the sum.

You can use any combination of these row operations to get a given matrix into row echelon form. The reduced row echelon form takes more steps but shows your solutions without further computation.

The focus is on the forms and the row operations in this section. To really dig in deep and discover how to get a matrix in row echelon form, read on to the next section.

EXAMPLE

Q. Write the system of equations
$$\begin{cases} 3x - y = 6 \\ 2x + 3y = 3 \end{cases}$$ as a matrix system.

A. $\begin{bmatrix} 3 & -1 \\ 2 & 3 \end{bmatrix} \begin{bmatrix} x \\ y \end{bmatrix} = \begin{bmatrix} 6 \\ 3 \end{bmatrix}$. The matrix on the left is the coefficient matrix, containing all the coefficients from the system. The second matrix is the variable matrix, and the third one, on the right, is the answer matrix. This completes the job for this question.

Q. Write the system from the preceding question as an augmented matrix.

A. $\left[\begin{array}{cc|c} 3 & -1 & 6 \\ 2 & 3 & 3 \end{array}\right]$. Just take the coefficient matrix and add on the answer matrix. You have an augmented matrix. Notice that the vertical line separates the two original matrices and lets you know that this is an augmented matrix and not a normal 2×3 matrix.

27 Using the augmented matrix from the last example, use elementary row operations to find $-3r_2 \rightarrow r_2$.

28 Now, using your answer from Problem 27, find $r_1 \leftrightarrow r_2$.

29 Now, keep going and find $r_1 + r_2 \rightarrow r_2$.

30 Lastly, find $3r_2 + r_1 \rightarrow r_1$.

Solving Systems of Equations Using Matrices

You can solve a system of equations using matrices in three ways. Putting a matrix in row echelon form using the techniques described in the last section is called *Gaussian elimination*. The second way uses a method called *inverse matrices*, and the third method is called *Cramer's Rule*. You may want to use one, two, or all three of these methods, depending on the situation. All three are described in the following sections.

Gaussian elimination

The process of putting a matrix in row echelon form is called *Gaussian elimination*. The focus in this section is on matrices in augmented form because that's most commonly what you'll be asked to do, but know that the rules don't change if you're asked to do this with some other form of matrix. The goals of using the elementary row operations are simple: Get a 1 in the upper-left corner of the matrix, get 0s in all positions underneath this 1, get 1s for all leading coefficients diagonally from the upper-left to the lower-right corners, and then get 0s below each of them. When you get to that point, you use a process called *back substitution* to solve for all the variables in the system.

Q. Put the system of equations $\begin{cases} 3x - y = 6 \\ 2x + 3y = 3 \end{cases}$ in augmented form; then write the matrix in row echelon form and solve the system.

A. $\begin{bmatrix} 1 & -\dfrac{1}{3} & 2 \\ 0 & \dfrac{11}{3} & -1 \end{bmatrix}$, $x = \dfrac{21}{11}$, $y = -\dfrac{3}{11}$.

In the last section, you see this system as an augmented matrix: $\begin{bmatrix} 3 & -1 & 6 \\ 2 & 3 & 3 \end{bmatrix}$.

Now you need to get it into row echelon form. First, you need to get a 1 in the upper-left corner. Follow the elementary row operation $\frac{1}{3}r_1 \to r_1$ to get $\begin{bmatrix} 1 & -\dfrac{1}{3} & 2 \\ 2 & 3 & 3 \end{bmatrix}$. Now, to make the

first element of row two 0, you need to add −2 to the current element. So use $-2r_1 + r_2 \to r_2$ to get $\begin{bmatrix} 1 & -\dfrac{1}{3} & 2 \\ 0 & \dfrac{11}{3} & -1 \end{bmatrix}$. This gives you an equation in the second row that's easy to solve for y: $\frac{11}{3}y = -1$, or $y = -\frac{3}{11}$. Now you can work backward using back substitution in the top equation: $x - \frac{1}{3}y = 2$.

You know the value of y, so substitute that in and get $x - \frac{1}{3}\left(-\frac{3}{11}\right) = 2$, or $x = \frac{21}{11}$.

---

31 Solve the system of equations $\begin{cases} 2x + 5y = 7 \\ 3x - 5y = 2 \end{cases}$ by writing it in augmented form and then putting the matrix in row echelon form.

32 Use Gaussian elimination to solve
$$\begin{cases} 3x - 2y + 6z = 7 \\ x - 2y - z = -2 \\ -3x + 10y + 11z = 18 \end{cases}$$

Inverse matrices

Another way to solve a system of linear equations is by using an *inverse matrix*. This process is based on the idea that if you write a system in matrix form, you'll have the coefficient matrix multiplying the variable matrix on the left side; and, if only you could divide a matrix, you'd have it made in the shade! Well, if you look at the simple equation $3x = 12$, you can solve it by dividing both sides by 3 or multiplying both sides by $\frac{1}{3}$, to get $x = 4$. This process of multiplying by a multiplicative inverse on both sides turns out to be a technique used in matrices! You have to use an inverse matrix. In Chapter 3 you see that if $f(x)$ is a function, its inverse is denoted by $f^{-1}(x)$. This is true for matrices as well: If A is the matrix, A^{-1} is its inverse. If you have three matrices (A, B, and C) and you know that AB = C, then you can solve for B by multiplying the inverse matrix A^{-1} on both sides:

$A^{-1}[AB] = A^{-1}C$, which simplifies to $B = A^{-1}C$.

Finding a matrix's inverse

TIP

But how do you find a matrix's inverse? Realize first that only square matrices have inverses. The number of rows must be equal to the number of columns. Even then, not every square matrix has an inverse. If the determinant (which is covered in the next section) of a matrix is 0, it doesn't have an inverse. And, getting right to it, here's how to find the determinant of a 2×2 matrix:

$$\begin{vmatrix} a & b \\ c & d \end{vmatrix} = ad - bc$$

where $\begin{vmatrix} a & b \\ c & d \end{vmatrix}$ denotes the determinant of the matrix $\begin{bmatrix} a & b \\ c & d \end{bmatrix}$. When a matrix does have an inverse, you can use several ways to find it depending on how big the matrix is. If it's a 2×2, you can find it by hand using the simple formula given here. If it's 3×3 or bigger, you *can* find it by hand, but it will be much more involved. This is where scientific calculators come in very handy.

In the meantime, if matrix A is the 2×2 matrix $\begin{bmatrix} a & b \\ c & d \end{bmatrix}$, then its inverse, $\begin{bmatrix} a & b \\ c & d \end{bmatrix}^{-1}$, is found using $\begin{bmatrix} a & b \\ c & d \end{bmatrix}^{-1} = \frac{1}{ad - bc}\begin{bmatrix} d & -b \\ -c & a \end{bmatrix}$, provided that $ad - bc \neq 0$.

Using an inverse matrix to solve a system of linear equations

Now that you can find the inverse matrix, all you have to do to solve a system of equations is follow these steps:

1. **Write the system as a matrix equation.**

2. **Create the inverse matrix.**

3. **Multiply this inverse in front of both sides of the equation.**

4. **Cancel on the left side; multiply the matrices on the right.**

5. **Multiply the scalar.**

EXAMPLE

Q. Set up the matrix equation for the system $\begin{cases} 3x - 2y = -1 \\ x + y = 3 \end{cases}$ by using inverse matrices.

A. $x = 1, y = 2$. First, set up the matrix equation: $\begin{bmatrix} 3 & -2 \\ 1 & 1 \end{bmatrix}\begin{bmatrix} x \\ y \end{bmatrix} = \begin{bmatrix} -1 \\ 3 \end{bmatrix}$. Now, find the inverse matrix using the formula just given in this chapter:

$$\begin{bmatrix} 3 & -2 \\ 1 & 1 \end{bmatrix}^{-1} = \frac{1}{3(1) - (-2)(1)}\begin{bmatrix} 1 & 2 \\ -1 & 3 \end{bmatrix} = \frac{1}{5}\begin{bmatrix} 1 & 2 \\ -1 & 3 \end{bmatrix}.$$

Now, multiply this inverse on the left of both sides of the matrix equation:
$\frac{1}{5}\begin{bmatrix} 1 & 2 \\ -1 & 3 \end{bmatrix}\begin{bmatrix} 3 & -2 \\ 1 & 1 \end{bmatrix}\begin{bmatrix} x \\ y \end{bmatrix} = \frac{1}{5}\begin{bmatrix} 1 & 2 \\ -1 & 3 \end{bmatrix}\begin{bmatrix} -1 \\ 3 \end{bmatrix}$. To multiply a matrix by its inverse cancels everything on the left except for the variable matrix:

 $\frac{1}{5}\begin{bmatrix} 1 & 2 \\ -1 & 3 \end{bmatrix}\begin{bmatrix} 3 & -2 \\ 1 & 1 \end{bmatrix}\begin{bmatrix} x \\ y \end{bmatrix} = \frac{1}{5}\begin{bmatrix} 1 & 2 \\ -1 & 3 \end{bmatrix}\begin{bmatrix} -1 \\ 3 \end{bmatrix}$. That means all you have to do is multiply the matrices on the right and then multiply the scalar:

$\begin{bmatrix} x \\ y \end{bmatrix} = \frac{1}{5}\begin{bmatrix} 1 & 2 \\ -1 & 3 \end{bmatrix}\begin{bmatrix} -1 \\ 3 \end{bmatrix} = \frac{1}{5}\begin{bmatrix} -1+6 \\ 1+9 \end{bmatrix} = \frac{1}{5}\begin{bmatrix} 5 \\ 10 \end{bmatrix} = \begin{bmatrix} 1 \\ 2 \end{bmatrix}$. This gives the solutions from top to bottom as $x = 1, y = 2$.

---

33 Solve the system $\begin{cases} 4x - y = -10 \\ 2x + 3y = 16 \end{cases}$ using the inverse matrix.

34 Solve the system $\begin{cases} 4x + 3y = 17 \\ 2x - y = 11 \end{cases}$ using the inverse matrix.

Cramer's Rule

Cramer's Rule is a method based on determinants of matrices that's used to solve systems of equations. The determinant of a 2×2 matrix $\begin{bmatrix} a & b \\ c & d \end{bmatrix}$ is $ad - bc$. The determinant of a 3×3 matrix is found using a process called *diagonals*. If $\begin{bmatrix} a_1 & b_1 & c_1 \\ a_2 & b_2 & c_2 \\ a_3 & b_3 & c_3 \end{bmatrix}$, then first rewrite the first two columns immediately following the third. Draw three diagonal lines from the upper left to the lower right and three diagonal lines from the lower left to the upper right, as shown in Figure 13-1.

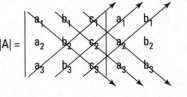

FIGURE 13-1: How to find a 3×3 matrix's determinant.

Then multiply down the three diagonals from left to right and up the other three. Find the sum of the products on the top and the sum of the products on the bottom. Finally, find the difference of the top and bottom. This is the same thing as $\left(a_1 b_2 c_3 + b_1 c_2 a_3 + c_1 a_2 b_3\right) - \left(a_3 b_2 c_1 + b_3 c_2 a_1 + c_3 a_2 b_1\right)$.

Using Cramer's Rule to find the solution of the 2×2 system, $\begin{cases} ax + by = c \\ dx + ey = f \end{cases}$:

$$x = \frac{\begin{vmatrix} c & b \\ f & e \end{vmatrix}}{\begin{vmatrix} a & b \\ d & e \end{vmatrix}} \qquad y = \frac{\begin{vmatrix} a & c \\ d & f \end{vmatrix}}{\begin{vmatrix} a & b \\ d & e \end{vmatrix}}$$

Observe that the numerator for x is simply the determinant of the matrix resulting from replacing the x coefficient column with the answer column in the coefficient matrix. The numerator for y is simply the determinant of the matrix resulting from replacing the y coefficient column with the answer column in the coefficient matrix.

EXAMPLE

Q. Solve the system of equations

$\begin{cases} -2x + 3y = 17 \\ -5x - y = 17 \end{cases}$ using Cramer's Rule.

A. $x = -4, y = 3$. Using Cramer's Rule by substituting the coefficients and the constants, $x = \dfrac{\begin{vmatrix} 17 & 3 \\ 17 & -1 \end{vmatrix}}{\begin{vmatrix} -2 & 3 \\ -5 & -1 \end{vmatrix}}$. Finding the determinants in the numerator and the denominator: $x = \dfrac{-17 - 51}{2 + 15} = \dfrac{-68}{17} = -4$

Do the same thing for

$$y = \frac{\begin{vmatrix} -2 & 17 \\ -5 & 17 \end{vmatrix}}{\begin{vmatrix} -2 & 3 \\ -5 & -1 \end{vmatrix}} = \frac{-34 + 85}{2 + 15} = \frac{51}{17} = 3.$$

Cramer's Rule is especially nice when the system of equations involves fractions or lots of prime numbers.

Find the determinant of $\begin{bmatrix} 2 & -1 & 4 \\ -3 & 4 & 6 \\ -2 & -1 & 5 \end{bmatrix}$

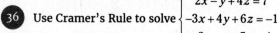 Use Cramer's Rule to solve $\begin{cases} 2x - y + 4z = 7 \\ -3x + 4y + 6z = -1 \\ -2x - y + 5z = 4 \end{cases}$

Answers to Problems on Systems of Equations

Following are particular answers to problems dealing with systems of equations. There is also guidance on getting the answers.

(1) Use substitution to solve the system $\begin{cases} x+y=6 \\ y=13-2x \end{cases}$. The answer is $x=7$ and $y=-1$.

Substituting the fact that $y=13-2x$, change the first equation to, $x+13-2x=6$. This simplifies to $-x=-7$, which means that $x=7$. Now that you know this value, you can substitute it into the second equation: $y=13-2(7)=13-14=-1$. The final answer: $x=7, y=-1$.

(2) The sum of two numbers is 14 and their difference is 2. Find the numbers. The answer is $x=8$ and $y=6$.

First, you need to change the given words into a system of equations using variables. The sum of two numbers being 14 becomes $x+y=14$, and their difference of 2 becomes $x-y=2$. The first equation has an x variable with a coefficient of 1, so you can solve for x easily getting: $x=14-y$. Now, substitute this expression for x in the other equation: $14-y-y=2$. Combine like terms and solve for y: $-2y=-12$ or $y=6$. Then substitute it into the other equation to solve for x: $x=16-6=8$.

(3) Solve the system $\begin{cases} \dfrac{x}{2}-\dfrac{y}{3}=-3 \\ \dfrac{2x}{3}+\dfrac{y}{2}=-2 \end{cases}$. The answer is $(x,y)=\left(-\dfrac{78}{17},\dfrac{36}{17}\right)$.

Sometimes the answers to these questions just aren't pretty. The first thing to do in any equation of this type is to get rid of the fractions by multiplying every term by the LCD. The LCD for both equations in this problem turns out to be 6, so, multiplying each term by 6 you get: $\begin{cases} 3x-2y=-18 \\ 4x+3y=-12 \end{cases}$. Which variable would you like to eliminate? How about y, because the terms already have opposite signs. Multiply the top equation by 3 and the bottom by 2: $\begin{cases} 9x-6y=-54 \\ 8x+6y=-24 \end{cases}$. Adding these two equations eliminates y: $17x=-78$, which means that $x=-\dfrac{78}{17}$. Substitute this value into the first equation to solve for y: $3\left(-\dfrac{78}{17}\right)-2y=-18$, $-\dfrac{234}{17}-2y=-18$, $-2y=-18+\dfrac{234}{17}=-\dfrac{72}{17}$ or $y=\dfrac{36}{17}$.

(4) Solve the system $\begin{cases} 3x-2y=4 \\ 6x-4y=8 \end{cases}$. The answer is $x=k$, $y=\dfrac{3k-4}{2}$.

You haven't seen this type of answer yet, in this book, so you need to see how this comes about. Notice first of all that all you have to do is multiply the top equation by -2 to get $-6x+4y=-8$, which is the exact opposite of the bottom equation, $6x-4y=8$. If you add these two equations together, you get $0=0$, which is always true. Therefore, this system has infinitely many solutions. Lots of answers will work in this system (actually, an infinite number of them). If you graph this system on a coordinate plane, you get two lines that lie on top of each other. How many points do those two lines share in common? All of them. If you want to write this out using variables to represent constants, then you can arbitrarily pick that $x=k$. When you plug that into the top equation to get $3k-2y=4$. Solve for y, and you get $y=\dfrac{3k-4}{2}$. If you had chosen to let $y=k$, then you would get that $x=\dfrac{2k+4}{3}$. It's your choice.

(5) Solve $\begin{cases} x^2 - y = 1 \\ x + y = 5 \end{cases}$. The answer is $x = 2, y = 3$ or $x = -3, y = 8$.

In this problem, you have a parabola and a line, so there is the possibility of two solutions. First, solve the linear equation for a variable, like x in the second equation: $x = 5 - y$. Now substitute this into the first equation: $(5 - y)^2 - y = 1$. Multiply out and combine like terms: $25 - 11y + y^2 = 1$. Then set the equation equal to 0: $24 - 11y + y^2 = 0$. This factors to $(3 - y)(8 - y) = 0$, which, when you use the zero product property, gets you two solutions for y: $y = 3$ and $y = 8$. Substitute them, one at a time, into the original quadratic equation to get the solutions for x. First: If $y = 3$, then $x^2 - 3 = 1$; $x^2 = 4$; $x = \pm 2$. When checking, you see that $2 + 3 = 5$ works in the second equation, but notice that $-2 + 3 \neq 5$ doesn't. That means that when y is 3, x only equals 2. Now do the same thing for $y = 8$: $x^2 - 8 = 1$; $x^2 = 9$; $x = \pm 3$. In the second equation, $3 + 8 \neq 5$ is false, but $-3 + 8 = 5$ is true, so the other solution is $x = -3, y = 8$.

(6) Solve the system of equations $\begin{cases} x + y = 9 \\ xy = 20 \end{cases}$. The answer is $x = 4, y = 5$ or $x = 5, y = 4$.

The intersection of a line and a hyperbola has the possibility of two solutions. First, solve the linear equation for x: $x = 9 - y$. Plug this into the second given equation: $(9 - y)y = 20$. Distribute to get $9y - y^2 = 20$ and rewrite as: $0 = y^2 - 9y + 20$, which factors to $0 = (y - 5)(y - 4)$. This means that y is 5 or 4. Plug them both into the linear equation. When $y = 5$: $x + 5 = 9$, giving you $x = 4$. And when $y = 4$: $x + 4 = 9$, or $x = 5$.

(7) Solve $\begin{cases} x^2 + y^2 = 1 \\ x + y^2 = -5 \end{cases}$. The answer is that it has no real solutions but does have the complex solutions: $\left(-2, \pm i\sqrt{3}\right)$ or $\left(3, \pm 2i\sqrt{2}\right)$.

Here, you have a circle and a parabola. There's the possibility of as many as four intersections or solutions. First, notice that both given equations have y^2 in them, with the same signs.

If you multiply the second equation by -1, you get $\begin{cases} x^2 + y^2 = 1 \\ -x - y^2 = 5 \end{cases}$. Adding the equations together to get $x^2 - x = 6$. Next, writing the equation as $x^2 - x - 6 = 0$. This factors to $(x - 3)(x + 2) = 0$, which does give two solutions, $x = 3$ or $x = -2$. When you plug $x = 3$ into the original second equation, you obtain $-3 - y^2 = 5$, which is equal to $y^2 = -8$; the square of a real number can't be negative, so this doesn't give you any real solutions. When you plug $x = -2$ into that same equation, you obtain $2 - y^2 = 5$ or $y^2 = -3$. Again, no real solutions. If you're open to complex solutions, then solving $y^2 = -8$ for y, you get $y = \pm 2i\sqrt{2}$, and when $y^2 = -3$, you have $y = \pm i\sqrt{3}$.

(8) Solve $\begin{cases} 27x^2 - 16y^2 = -400 \\ 4y^2 - 9x^2 = 36 \end{cases}$. The answer is $\left(\pm\dfrac{16}{3}, \pm\sqrt{73}\right)$ or $\left(\pm\dfrac{16}{3}, \mp\sqrt{73}\right)$.

The two hyperbolas can intersect in as many as four places. The y terms have opposite signs, so it's easier to eliminate them after you multiply the second equation by 4: $\begin{cases} 27x^2 - 16y^2 = -400 \\ -36x^2 + 16y^2 = 144 \end{cases}$. Adding these two equations gets you $-9x^2 = -256$, or $x^2 = \dfrac{256}{9}$, which finally means that $x = \pm\dfrac{16}{3}$. Now notice that both of the original equations have x^2 in them, but no x term. If you square $\dfrac{16}{3}$ or $-\dfrac{16}{3}$, you get the same result: $\dfrac{256}{9}$. This means the positive and negative signs don't really matter when it comes to solving for y. Using the first equation: $27\left(\dfrac{256}{9}\right) - 16y^2 = -400$. Next, simplify: $768 - 16y^2 = -400$ to get $-16y^2 = -1{,}168$ or $y^2 = 73$. That gives you $y = \pm\sqrt{73}$. So the four solutions are: $\left(\dfrac{16}{3}, \sqrt{73}\right), \left(\dfrac{16}{3}, -\sqrt{73}\right), \left(-\dfrac{16}{3}, \sqrt{73}\right)$ and $\left(-\dfrac{16}{3}, -\sqrt{73}\right)$.

9 Solve $\begin{cases} \dfrac{14}{x+3} + \dfrac{7}{4-y} = 9 \\ \dfrac{21}{x+3} - \dfrac{3}{4-y} = 0 \end{cases}$. The solution is $x = 4$ and $y = 3$.

First, rewrite the system by letting $u = \dfrac{1}{x+3}$ and $v = \dfrac{1}{4-y}$ and getting $\begin{cases} 14u + 7v = 9 \\ 21u - 3v = 0 \end{cases}$. Now,

multiply the first equation by 3 and the second equation by 7: $\begin{cases} 42u + 21v = 27 \\ 147u - 21v = 0 \end{cases}$. Adding these

two equations gets you $189u = 27$, which means that $u = \dfrac{1}{7}$. Remember, though, you're looking

for x, not u. So, going back to $u = \dfrac{1}{x+3}$, write it as $\dfrac{1}{7} = \dfrac{1}{x+3}$ and you get that $x = 4$. Now you

can use that to get $y = 3$, starting with the top equation in u and v: $14u + 7v = 9$ becomes

$14\left(\dfrac{1}{7}\right) + 7v = 9$ or $2 + 7v = 9$. The value of v is 1, so go back to $v = \dfrac{1}{4-y}$ and solve for y. $\dfrac{1}{1} = \dfrac{1}{4-y}$

gives you that $1 = 4 - y$ or $y = 3$.

Because this is a rational expression, also be sure to always check your solution to see whether it's extraneous. In other words, if $x = 4$ or $y = 3$, do you get 0 in the denominator of either given equation? In this case, the answer is no — so this solution is legit!

10 Solve $\begin{cases} \dfrac{12}{x+1} - \dfrac{12}{y-1} = 8 \\ \dfrac{6}{x+1} + \dfrac{6}{y-1} = -2 \end{cases}$. The answer is $x = 5$ and $y = -1$.

If you let $u = \dfrac{1}{x+1}$ and $v = \dfrac{1}{y-1}$, you can rewrite the system as $\begin{cases} 12u - 12v = 8 \\ 6u + 6v = -2 \end{cases}$. Now, multiply

the second equation by 2 and get $\begin{cases} 12u - 12v = 8 \\ 12u + 12v = -24 \end{cases}$. Add them to get $24u = 4$, or $u = \dfrac{1}{6}$. Work

your way backward from there: $u = \dfrac{1}{x+1}$ means $\dfrac{1}{6} = \dfrac{1}{x+1}$ or $x = 5$. Using the adjusted first

equation, $12u - 12v = 8$, you have $12\left(\dfrac{1}{6}\right) - 12v = 8$ or $2 - 12v = 8$ which becomes $-12v = 6$ or $v = -\dfrac{1}{2}$.

Since $v = \dfrac{1}{y-1}$, you write $\dfrac{1}{-2} = \dfrac{1}{y-1}$, and when $-2 = y - 1$ you get $y = -1$.

11 Solve the system $\begin{cases} 3x - 2y = 17 \\ x - 2z = 1 \\ 3y + 2z = 1 \end{cases}$. The answer is $(x, y, z) = (5, -1, 2)$.

Each equation has just two of the three variables. The second and third equations have z variables with opposite coefficients, so add them together to get $x + 3y = 2$. Use this new equation and the first equation in the next step. Multiply the new equation through by -3,

and you have $\begin{cases} 3x - 2y = 17 \\ -3x - 9y = -6 \end{cases}$. Adding the two equations together, you get $-11y = 11$ or $y = -1$.

Substituting back into the original first equation, you get $3x - 2(-1) = 17$ or $x = 5$. Finally, going back to the original second equation, let $x = 5$ and you get $5 - 2z = 1$ or $z = 2$.

12 Solve $\begin{cases} 2x - y + z = 1 \\ x + y - z = 2 \\ -x - y + z = 2 \end{cases}$. The answer is no solution.

Right away, notice that all the coefficients on the middle and bottom equations are exact opposites of each other. When you add these two equations you get $0 = 4$. Because this equation is false, there's no solution.

13 Solve $\begin{cases} 2x+3y+4z=37 \\ 4x-3y+2z=17 \\ x+2y-3z=-5 \\ 3x-2y+z=11 \end{cases}$. The answer is $x=4$, $y=3$, and $z=5$.

The first and second equations have y terms with opposite coefficients, and so do the third and fourth equations. Add the pairs together to get two new equations in just x and z.

$$\begin{cases} 6x+6z=54 \\ 4x-2z=6 \end{cases}$$

Multiply the second equation through by 3, and then add the two equations together.

$$\begin{cases} 6x+6z=54 \\ 12x-6z=18 \end{cases}$$

$18x=72$ or $x=4$. Substituting back into the first equation in x and z, you have $6(4)+6z=54$, giving you $6z=30$ or $z=5$. Finally, using the original first equation to solve for y, $2(4)+3y+4(5)=37$ giving you $3y=9$ or $y=3$.

Note: It's not "typical" to have a solution in cases when a system has more equations than variables. It happens only because this system happens to be redundant rather than inconsistent.

14 Solve $\begin{cases} 3a+b+c+d=0 \\ 4a+5b+2c=15 \\ 4a+2b+5d=-10 \\ -5a+3b-d=8 \end{cases}$. The answer is $a=1$, $b=3$, $c=-2$, $d=-4$.

Look for a good candidate to eliminate first. There's no d term in the second equation, and the d terms are opposites in the first and last. There's where you can start.

Add the first and last equations together, and add 5 times the first equation to the third equation. Your new system, without the d term, is:

original second equation $\left\{\begin{array}{l} 4a+5b+2c=15 \end{array}\right.$
first plus last equation $ -2a+4b+c=8$
−5 times first plus third $\left. -11a-3b-5c=-10 \right.$

The next good candidate for elimination is the c term. Use multiples of the second term in the new system to do the elimination.

−2 times second plus first $\left\{\begin{array}{l} 8a-3b=-1 \end{array}\right.$
5 times second plus third $\left. -21a+17b=30 \right.$

Solving for a or b will require multiplying both equations by some number to make the coefficients the opposite. If you multiply the first new equation by 17 and the second by 3, you get:

$$\begin{cases} 136a-51b=-17 \\ -63a+51b=90 \end{cases}$$

Adding the equations together, you have $73a = 73$ or $a = 1$. Back-substituting this into $8a - 3b = -1$, you get $b = 3$. Then, using $-2a + 4b + c = 8$ and the two values you've found, you get that $c = -2$. And, finally, if $3a + b + c + d = 0$, you have $3(1) + 3 + (-2) + d = 0$, giving you $d = -4$.

(15) Sketch the graph of $\begin{cases} 2x + y \geq 9 \\ 2x - y \geq 1 \\ x \leq 7 \end{cases}$. See the graph for the answer.

Put the first two inequalities in slope-intercept form first. The top inequality is $y \geq -2x + 9$; the second inequality is $y \leq 2x - 1$. Put these inequalities and $x \leq 7$ on the same graph.

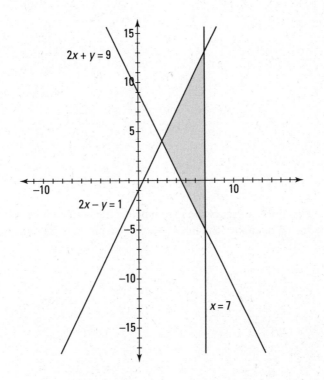

(16) Sketch the graph of $\begin{cases} x^2 + y^2 \geq 9 \\ x^2 + (y-3)^2 \geq 9 \end{cases}$. See the graph for the answer.

Both of these inequalities describe regions bounded by circles. The answer is the area outside of both these two circles. If you don't recognize them as such, turn to Chapter 12 to read up on conic sections. Graph them both on the same graph.

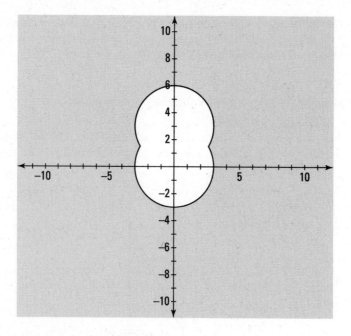

(17) Sketch the graph of $\begin{cases} x^2 - y > 2 \\ x - y < 4 \end{cases}$. See the graph for the answer.

The first inequality represents the region below the parabola, $y = x^2 - 2$. The second inequality represents the region above the line, $y = x - 4$. Graph them both on the same graph.

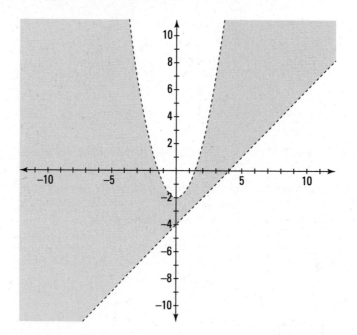

18 Sketch the graph of $\begin{cases} y \ge 0 \\ x + y < 4 \\ y \le \sqrt{x-1} \end{cases}$. See the graph for the answer.

The expression in the third inequality is a square root function. If you don't remember how to graph it, turn to Chapter 3 for a refresher.

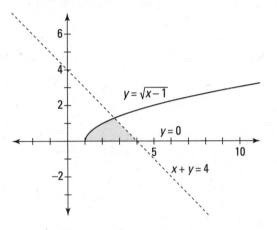

19 Find the constants A and B: $\dfrac{x-38}{x^2+x-12} = \dfrac{A}{x+4} + \dfrac{B}{x-3}$. The answer is $A = 6$ and $B = -5$.

The decomposition process has been started for you, showing the factorization of the denominator of the fraction being decomposed. The first thing you should do is multiply every term in the equation by the factored denominator

$\dfrac{x-38}{x^2+x-12} \cdot (x+4)(x-3) = \dfrac{A}{x+4} \cdot (x+4)(x-3) + \dfrac{B}{x-3} \cdot (x+4)(x-3)$; after reducing the fractions, you have:

$x - 38 = A(x-3) + B(x+4)$. Multiply everything out to get $x - 38 = Ax - 3A + Bx + 4B$. Collect the like terms on the right: $x - 38 = A(x-3) + B(x+4)$ or $x - 38 = Ax - 3A + Bx + 4B$. Now factor out the x on the right: $x - 38 = (A+B)x - 3A + 4B$. The coefficients of the x terms are equal, which gives you one equation: $1 = A + B$. The constants are also equal, which gives you a second equation: $-38 = -3A + 4B$. Solving this new system of equations, $\begin{cases} A+B = 1 \\ -3A+4B = -38 \end{cases}$, you first multiply the top equation by 3 and then add the two equations together to get $7B = -35$. This gives you that $B = -5$. Substituting this into the top equation, $A + (-5) = 1$ gives you that $A = 6$.

20 Find the initial form of the partial fraction decomposition for $\dfrac{5x-4}{(x-1)^2}$ (but don't solve for the constants).

The answer is $\dfrac{5x-4}{(x-1)^2} = \dfrac{A}{(x-1)^2} + \dfrac{B}{x-1}$.

This is as far as you need to go. You just need to set up the problem, not to perform the decomposition. The denominator is already factored for you. Just remember to create one term for every power of this binomial, up to its degree of 2. In other words, you need one fraction with $(x-1)^2$ in it, and you need another with $(x-1)^1$ in it.

(21) Find the partial fraction decomposition for $\dfrac{2x^2-21x+18}{(x-1)(x^2-4x+4)}$. The answer is

$$\dfrac{2x^2-21x+18}{(x-1)(x^2-4x+4)} = -\dfrac{1}{x-1} - \dfrac{16}{(x-2)^2} + \dfrac{3}{x-2}.$$

First, factor the given denominator to get $\dfrac{2x^2-21x+18}{(x-1)(x-2)^2}$. Because each factor is linear, set up

three different fractions with constants on the top: one for the $(x-1)$ factor, one for the

$(x-2)^2$ factor, and one for the $(x-2)$ factor: $\dfrac{2x^2-21x+18}{(x-1)(x-2)^2} = \dfrac{A}{x-1} + \dfrac{B}{(x-2)^2} + \dfrac{C}{x-2}$.

Multiply every term by the factored denominator, cancel, and get $2x^2-21x+18 = A(x-2)^2 + B(x-1) + C(x-1)(x-2)$.

Multiply out and collect the like terms: $2x^2-21x+18 = Ax^2 + Cx^2 - 4Ax + Bx - 3Cx + 4A - B + 2C$.
Factor out the x^2 and the x: $2x^2-21x+18 = (A+C)x^2 + (-4A+B-3C)x + 4A-B+2C$.

Set the coefficients of x^2 equal to each other: $2 = A+C$. Set the coefficients of x equal to each other: $-21 = -4A+B-3C$. Lastly, set the constants equal to each other: $18 = 4A-B+2C$. Solve this system of equations using elimination to get $A=-1$, $B=-16$, and $C=3$. This is where you get the answer $\dfrac{2x^2-21x+18}{(x-1)(x-2)^2} = \dfrac{A}{x-1} + \dfrac{B}{(x-2)^2} + \dfrac{C}{x-2} = \dfrac{-1}{x-1} + \dfrac{-16}{(x-2)^2} + \dfrac{3}{x-2}$.

(22) Find the partial fraction decomposition for $\dfrac{11x^2-7x+14}{2x^3-4x^2+3x-6}$. The answer is

$$\dfrac{11x^2-7x+14}{2x^3-4x^2+3x-6} = \dfrac{3x-1}{2x^2+3} + \dfrac{4}{x-2}.$$

Factor the denominator by using grouping: $2x^2(x-2)+3(x-2)$ becomes $(x-2)(2x^2+3)$.
Use this to make two different fractions with a constant on top of the linear factor and a linear expression on top of the quadratic factor: $\dfrac{11x^2-7x+14}{2x^3-4x^2+3x-6} = \dfrac{Ax+B}{2x^2+3} + \dfrac{C}{x-2}$. Multiply each

term by the factored denominator, cancel, and get $11x^2-7x+14 = (Ax+B)(x-2) + C(2x^2+3)$.
Multiply it all out and collect the like terms and factor to get
$11x^2-7x+14 = (A+2C)x^2 + (-2A+B)x - 2B + 3C$. This gives you a system with three

equations: $\begin{cases} A+2C = 11 \\ -2A+B = -7 \\ -2B+3C = 14 \end{cases}$. Solving this system tells you that $A=3$, $B=-1$, and $C=4$, which

gives you the answer $\dfrac{11x^2-7x+14}{2x^3-4x^2+3x-6} = \dfrac{3x-1}{2x^2+3} + \dfrac{4}{x-2}$.

(23) Find $4D$. The answer is $4D = \begin{bmatrix} 8 & 16 & 20 & -32 \\ 40 & 12 & -8 & -12 \end{bmatrix}$

First, write out the problem: $4D = 4\begin{bmatrix} 2 & 4 & 5 & -8 \\ 10 & 3 & -2 & -3 \end{bmatrix}$

Distribute the 4 to every element inside the matrix to get the answer: $\begin{bmatrix} 8 & 16 & 20 & -32 \\ 40 & 12 & -8 & -12 \end{bmatrix}$

(24) Find $4D + 5C$. The answer is $4D + 5C = \begin{bmatrix} -17 & 11 & 35 & -2 \\ 40 & 22 & -18 & 18 \end{bmatrix}$

First, substitute the given matrices into the expression:

$$4D + 5C = 4\begin{bmatrix} 2 & 4 & 5 & -8 \\ 10 & 3 & -2 & -3 \end{bmatrix} + 5\begin{bmatrix} -5 & -1 & 3 & 6 \\ 0 & 2 & -2 & 6 \end{bmatrix}$$

Distribute both scalars to every element of their matrix:

$$= \begin{bmatrix} 8 & 16 & 20 & -32 \\ 40 & 12 & -8 & -12 \end{bmatrix} + 5\begin{bmatrix} -25 & -5 & 15 & 30 \\ 0 & 10 & -10 & 30 \end{bmatrix}.$$ Add these two matrices by adding their

corresponding elements to get $\begin{bmatrix} -17 & 11 & 35 & -2 \\ 40 & 22 & -18 & 18 \end{bmatrix}$

(25) Find $3C - E$. The answer is no solution.

These matrices aren't the same dimensions, so you can't add them. There's no solution.

(26) Find DE. The answer is $DE = \begin{bmatrix} 64 & 19 & 11 \\ 13 & 20 & -9 \end{bmatrix}$

Substitute the given matrices into the expression $DE = \begin{bmatrix} 2 & 4 & 5 & -8 \\ 10 & 3 & -2 & -3 \end{bmatrix} \begin{bmatrix} -1 & 2 & -1 \\ 4 & 4 & 0 \\ 2 & 3 & 1 \\ -5 & 2 & -1 \end{bmatrix}$. Check to

see whether you can even multiply them. The matrix on the left is 2×4 and the one on the right is 4×3, so you can multiply them. Multiply every row from the left matrix by every column from the right matrix.

The sum of the first row times the first column: $2(-1) + 4(4) + 5(2) - 8(-5) = 64$.

The sum of the first row times the second column: $2(2) + 4(4) + 5(3) - 8(2) = 19$.

The sum of the first row times the third column: $2(-1) + 4(0) + 5(1) - 8(-1) = 11$.

The sum of the second row times the first column: $10(-1) + 3(4) - 2(2) - 3(-5) = 13$.

The sum of the second row times the second column: $10(-1) + 3(4) - 2(3) - 3(2) = 20$.

The sum of the second row times the third column: $10(-1) + 3(0) - 2(1) - 3(-1) = -9$.

Putting these all into a matrix gives you the answer: $DE = \begin{bmatrix} 64 & 19 & 11 \\ 13 & 20 & -9 \end{bmatrix}$

(27) Using the augmented matrix $\begin{bmatrix} 3 & -1 & | & 6 \\ 2 & 3 & | & 3 \end{bmatrix}$, use elementary row operations to find $-3r_2 \to r_2$.

The answer is $\begin{bmatrix} 3 & -1 & | & 6 \\ -6 & -9 & | & -9 \end{bmatrix}$.

Just multiply the second row of the given equation by -3 and replace that row with the

answers to get $\begin{bmatrix} 3 & -1 & | & 6 \\ 2 & 3 & | & 3 \end{bmatrix} -3r_2 \to r_2 \begin{bmatrix} 3 & -1 & | & 6 \\ -6 & -9 & | & -9 \end{bmatrix}$

(28) Now, using your answer from Problem 27, find $r_1 \leftrightarrow r_2$. The answer is $\begin{bmatrix} -6 & -9 & | & -9 \\ 3 & -1 & | & 6 \end{bmatrix}$

Swap the first row with the second row and you get $\begin{bmatrix} 3 & -1 & | & 6 \\ -6 & -9 & | & -9 \end{bmatrix} r_1 \leftrightarrow r_2 \begin{bmatrix} -6 & -9 & | & -9 \\ 3 & -1 & | & 6 \end{bmatrix}$.

29 Now, keep going and find $r_1 + r_2 \to r_2$. The answer is $\begin{bmatrix} -6 & -9 & | & -9 \\ -3 & -10 & | & -3 \end{bmatrix}$

Add each element from row one to its corresponding element in row two and put the results in row two: $\begin{bmatrix} -6 & -9 & | & -9 \\ 3 & -1 & | & 6 \end{bmatrix} r_1 + r_2 \to r_2 \begin{bmatrix} -6 & -9 & | & -9 \\ -3 & -10 & | & -3 \end{bmatrix}$

30 Lastly, find $3r_2 + r_1 \to r_1$. The answer is $\begin{bmatrix} -15 & -39 & | & -18 \\ -3 & -10 & | & -3 \end{bmatrix}$.

Temporarily multiply the second row by 3 to get $\begin{bmatrix} -9 & -30 & | & -9 \end{bmatrix}$. Add these to the corresponding elements in row one and replace row one with the results and you get

$\begin{bmatrix} -6 & -9 & | & -9 \\ -3 & -10 & | & -3 \end{bmatrix} 3r_2 + r_1 \to r_1 \begin{bmatrix} -15 & -39 & | & -18 \\ -3 & -10 & | & -3 \end{bmatrix}$

31 Solve the system of equations $\begin{cases} 2x + 5y = 7 \\ 3x - 5y = 2 \end{cases}$ by writing it in augmented form and then putting the matrix in row echelon form. The answer is $x = \dfrac{9}{5}, y = \dfrac{17}{25}$.

The matrix in augmented form is $\begin{bmatrix} 2 & 5 & | & 7 \\ 3 & -5 & | & 2 \end{bmatrix}$. Multiply the top row by $\dfrac{1}{2}$ to get a 1 in the upper-left corner: $\begin{bmatrix} 1 & \frac{5}{2} & | & \frac{7}{2} \\ 3 & -5 & | & 2 \end{bmatrix}$. To get a 0 under the 1 you just created, add -3 times the first row to the second row: $\begin{bmatrix} 1 & \frac{5}{2} & | & \frac{7}{2} \\ 3 & -5 & | & 2 \end{bmatrix} -3r_1 + r_2 \to r_2 \begin{bmatrix} 1 & \frac{5}{2} & | & \frac{7}{2} \\ 0 & -\frac{25}{2} & | & -\frac{17}{2} \end{bmatrix}$. Set up an equation from the

second row: $-\dfrac{25}{2}y = -\dfrac{17}{2}$. Solve this equation to get $y = -\dfrac{17}{2}\left(-\dfrac{2}{25}\right) = \dfrac{17}{25}$. Use that answer in an

equation created from the first row, $x + \dfrac{5}{2}\left(\dfrac{17}{25}\right) = \dfrac{7}{2}$ to get that $x = \dfrac{7}{2} - \dfrac{17}{10} = \dfrac{35 - 17}{10} = \dfrac{18}{10} = \dfrac{9}{5}$.

32 Use Gaussian elimination to solve $\begin{cases} 3x - 2y + 6z = 7 \\ x - 2y - z = -2 \\ -3x + 10y + 11z = 18 \end{cases}$. The answer is $x = 1, y = 1, z = 1$.

Set up the system as an augmented matrix: $\begin{bmatrix} 3 & -2 & 6 & | & 7 \\ 1 & -2 & -1 & | & -2 \\ -3 & 10 & 11 & | & 18 \end{bmatrix}$

$r_1 \leftrightarrow r_2$ gets a 1 in the upper-left corner: $= \begin{bmatrix} 1 & -2 & -1 & | & -2 \\ 3 & -2 & 6 & | & 7 \\ -3 & 10 & 11 & | & 18 \end{bmatrix}$

$-3r_1 + r_2 \to r_2$ gets a 0 under the 1 in the second row: $= \begin{bmatrix} 1 & -2 & -1 & | & -2 \\ 0 & 4 & 9 & | & 13 \\ -3 & 10 & 11 & | & 18 \end{bmatrix}$

$3r_1 + r_3 \to r_3$ gets a 0 under the 1 in the third row: $= \begin{bmatrix} 1 & -2 & -1 & | & -2 \\ 0 & 4 & 9 & | & 13 \\ 0 & 4 & 8 & | & 12 \end{bmatrix}$

Next, to avoid fractions, switch rows two and three, and then multiply the new row two by $\frac{1}{4}$.

$$r_2 \leftrightarrow r_3 = \begin{bmatrix} 1 & -2 & -1 & | & -2 \\ 0 & 4 & 8 & | & 12 \\ 0 & 4 & 9 & | & 13 \end{bmatrix} \text{ and then } \frac{1}{4}r_2 \rightarrow r_2 = \begin{bmatrix} 1 & -2 & -1 & | & -2 \\ 0 & 1 & 2 & | & 3 \\ 0 & 4 & 9 & | & 13 \end{bmatrix}$$

$-4r_2 + r_3 \rightarrow r_3$ gets a 0 under the 1 you just created on the main diagonal: $= \begin{bmatrix} 1 & -2 & -1 & | & -2 \\ 0 & 1 & 2 & | & 3 \\ 0 & 0 & 1 & | & 1 \end{bmatrix}$.

Reading the corresponding equation from the third row, you have that $z = 1$. Back substituting, you use the second row to get that $y + 2(1) = 3$ or $y = 1$. Back substitute again using the first row, and you have: $x - 2(1) - 1(1) = -2$ or $x = 1$.

(33) Solve the system $\begin{cases} 4x - y = -10 \\ 2x + 3y = 16 \end{cases}$ using inverse matrices. The answer is $x = -1$ and $y = 6$.

First, write the system as a matrix equation: $\begin{bmatrix} 4 & -1 \\ 2 & 3 \end{bmatrix} \begin{bmatrix} x \\ y \end{bmatrix} = \begin{bmatrix} -10 \\ 16 \end{bmatrix}$. Now, find the inverse

matrix using the handy formula found in the section on "Inverse matrices":

$$\frac{1}{4(3) - (-1)(2)} \begin{bmatrix} 3 & 1 \\ -2 & 4 \end{bmatrix} = \frac{1}{14} \begin{bmatrix} 3 & 1 \\ -2 & 4 \end{bmatrix}.$$

Multiply both sides of the matrix equation by the inverse:

$$\frac{1}{14} \begin{bmatrix} 3 & 1 \\ -2 & 4 \end{bmatrix} \begin{bmatrix} 4 & -1 \\ 2 & 3 \end{bmatrix} \begin{bmatrix} x \\ y \end{bmatrix} = \frac{1}{14} \begin{bmatrix} 3 & 1 \\ -2 & 4 \end{bmatrix} \begin{bmatrix} -10 \\ 16 \end{bmatrix} \text{ gives you } \begin{bmatrix} x \\ y \end{bmatrix} = \frac{1}{14} \begin{bmatrix} 3 & 1 \\ -2 & 4 \end{bmatrix} \begin{bmatrix} -10 \\ 16 \end{bmatrix}$$

Multiply the two matrices on the right, and then multiply by the scalar:

$\begin{bmatrix} x \\ y \end{bmatrix} = \frac{1}{14} \begin{bmatrix} -14 \\ 84 \end{bmatrix} = \begin{bmatrix} -1 \\ 6 \end{bmatrix}$. Your solutions from top to bottom are $x = -1$ and $y = 6$.

(34) Solve the system $\begin{cases} 4x + 3y = 17 \\ 2x - y = 11 \end{cases}$ using inverse matrices. The answer is $x = 5$ and $y = -1$.

Write the system as a matrix equation: $\begin{bmatrix} 4 & 3 \\ 2 & -1 \end{bmatrix} \begin{bmatrix} x \\ y \end{bmatrix} = \begin{bmatrix} 17 \\ 11 \end{bmatrix}$. Find the inverse:

$\frac{1}{4(-1) - (3)(2)} \begin{bmatrix} -1 & -3 \\ -2 & 4 \end{bmatrix} = \frac{1}{-10} \begin{bmatrix} -1 & -3 \\ -2 & 4 \end{bmatrix}$. Multiply this on both sides of the matrix equation:

$-\frac{1}{10} \begin{bmatrix} -1 & -3 \\ -2 & 4 \end{bmatrix} \begin{bmatrix} 4 & 3 \\ 2 & -1 \end{bmatrix} \begin{bmatrix} x \\ y \end{bmatrix} = -\frac{1}{10} \begin{bmatrix} -1 & -3 \\ -2 & 4 \end{bmatrix} \begin{bmatrix} 17 \\ 11 \end{bmatrix}$. Multiply the matrices: $\begin{bmatrix} x \\ y \end{bmatrix} = -\frac{1}{10} \begin{bmatrix} -50 \\ 10 \end{bmatrix}$

Multiply the scalar: $\begin{bmatrix} x \\ y \end{bmatrix} = \begin{bmatrix} 5 \\ -1 \end{bmatrix}$. You read that $x = 5$ and $y = -1$.

(35) Find the determinant of $\begin{bmatrix} 2 & -1 & 4 \\ -3 & 4 & 6 \\ -2 & -1 & 5 \end{bmatrix}$. The answer is 93.

Because this is a 3×3 matrix, you have to use diagonals. First, rewrite the first two columns

after the third one: $\begin{bmatrix} 2 & -1 & 4 \\ -3 & 4 & 6 \\ -2 & -1 & 5 \end{bmatrix} \begin{matrix} 2 & -1 \\ -3 & 4 \\ -2 & -1 \end{matrix}$. The sum of the diagonals from top-left to bottom-

right is: $(2)(4)(5) + (-1)(6)(-2) + 4(-3)(-1) = 40 + 12 + 12 = 64$. The sum of the diagonals from bottom-left to top-right is: $(-2)(4)(4) + (-1)(6)(2) + (5)(-3)(-1) = -32 - 12 + 15 = -29$. The difference of the first sum and the second sum is $64 - (-29) = 93$.

(36) Use Cramer's Rule to solve $\begin{cases} 2x - y + 4z = 7 \\ -3x + 4y + 6z = -1. \\ -2x - y + 5z = 4 \end{cases}$ The answer is $x = 1, y = -1, z = 1$.

Each variable will be determined by dividing its determinant by the coefficient determinant, $\begin{vmatrix} 2 & -1 & 4 \\ -3 & 4 & 6 \\ -2 & -1 & 5 \end{vmatrix}$.

Solve for x: $x = \dfrac{\begin{vmatrix} 7 & -1 & 4 \\ -1 & 4 & 6 \\ 4 & -1 & 5 \end{vmatrix}}{\begin{vmatrix} 2 & -1 & 4 \\ -3 & 4 & 6 \\ -2 & -1 & 5 \end{vmatrix}} = \dfrac{93}{93} = 1$

Solve for y: $y = \dfrac{\begin{vmatrix} 2 & 7 & 4 \\ -3 & -1 & 6 \\ -2 & 4 & 5 \end{vmatrix}}{\begin{vmatrix} 2 & -1 & 4 \\ -3 & 4 & 6 \\ -2 & -1 & 5 \end{vmatrix}} = \dfrac{-93}{93} = -1$

Solve for z: $z = \dfrac{\begin{vmatrix} 2 & -1 & 7 \\ -3 & 4 & -1 \\ -2 & -1 & 4 \end{vmatrix}}{\begin{vmatrix} 2 & -1 & 4 \\ -3 & 4 & 6 \\ -2 & -1 & 5 \end{vmatrix}} = \dfrac{93}{93} = -1$

Chapter **14**

Spotting Patterns in Sequences and Series

This chapter is all about patterns. No, it's not about making quilts, although there could be some overlap . . . nah! The patterns covered here are all about numbers, not cloth. Namely, you get good practice involving sequences, series, and the binomial theorem.

A *sequence* is an ordered list of numbers. A *series* is the sum of some of the terms in a sequence. Sequences and series often follow a pattern, and this is described with a mathematical formula. The *binomial theorem* is the result of discovering the pattern of an expanded binomial.

REMEMBER

One mathematical term that comes up frequently in this chapter is *factorial*, which has appeared in other mathematical subjects. The factorial operation, $n!$, read "n factorial," is defined $n! = 1 \cdot 2 \cdot 3 \cdots (n-1) \cdot n$. The n can't be negative. And, a special rule says that $0! = 1$.

General Sequences and Series: Determining Terms

Mathematically, the general notation for a sequence is written in the following form: $\{a_i\}_{i=1,\ldots,n} = a_1, a_2, a_3, \ldots, a_n$. Here, n is the number of terms, a_i is the i-th term of the sequence, a_1 is the first term, a_2 is the second term, and so on. Similarly, a series can be written as the sum of the terms: $a_1 + a_2 + a_3 + \cdots + a_n$. The pattern of sequences and series can usually be described by a general expression or rule. Because sequences and series can have any finite number of terms, this expression allows you to find any term in the list without having to find all the terms up to and including that particular term. If you're not given the rule for the general term, you can sometimes determine what it is if you're given the first few terms of a sequence or series.

TECHNICAL STUFF

Sometimes a term in a sequence depends on the term(s) before it. These are called *recursive sequences*. A famous example of a recursive sequence is the Fibonacci Sequence: $1, 1, 2, 3, 5, 8, 13, \ldots$ where each term is the sum of the two before it, after the first two terms are given.

EXAMPLE

Q. Write the first five terms of the sequence whose rule is $a_n = n^2 + 3$.

A. 4, 7, 12, 19, 28. To find each term, you just plug the number of the term (n) into the formula: $a_1 = (1)^2 + 3 = 4$; $a_2 = (2)^2 + 3 = 7$; $a_3 = (3)^2 + 3 = 12$; $a_4 = (4)^2 + 3 = 19$; $a_5 = (5)^2 + 3 = 28$.

Q. Write a general expression for the sequence to find the nth term if the first four terms are: $-\dfrac{1}{2}, \dfrac{2}{3}, -\dfrac{3}{4}, \dfrac{4}{5}$.

A. $a_n = (-1)^n \dfrac{n}{n+1}$. First, notice that the sign alternates between negative and positive. To deal with this, multiply by powers of -1: $(-1)^n$. Next, notice that the sequence's numerator is the same as the term number (n), so n becomes your numerator. Finally, you can see that the denominator is simply one number larger than the numerator (and term number), so it can be written as $n+1$. Putting these pieces together, you get $a_n = (-1)^n \dfrac{n}{n+1}$.

 1 Write the first five terms of the sequence whose rule is $a_n = \dfrac{n-3}{2n}$.

2 Write a general expression for the sequence to find the nth term: $2, 4, 10, 28, 82, \ldots$

3 Write a general expression for the sequence to find the nth term: $2, 2, \dfrac{8}{3}, 4, \dfrac{32}{5}, \ldots$

 4 Write the next two terms of the sequence $1, 1, 2, 3, 7, 16, \ldots$

 5 Find the sum of the first five terms of the series whose n-th term is described by $2^{n-1} + 1$.

 6 Find the sum of the first five terms of the series whose nth term is described by $3^n + 2n$.

Working Out the Common Difference: Arithmetic Sequences and Series

One special type of sequence is called an *arithmetic sequence*. In these sequences, each term differs from the one before it by a *common difference*, d. As a result, you have a formula for finding the nth term of an arithmetic sequence:

$$a_n = a_1 + (n-1)d$$

TECHNICAL STUFF where a_1 is the first term, n is the number of the term, and d is the common difference.

To find the sum of some terms of an arithmetic sequence, also called an *arithmetic series,* you have to add a given number of terms together. This can write in summation notation:

$$S_k = \sum_{n=1}^{k} a_n = \frac{k}{2}(a_1 + a_n)$$

This is read as "the *k*th partial sum of a_n" where $n = 1$ is the lower limit, *k* is the sum's upper limit, a_1 is the first term, and a_k is the last term to be added. To find a partial sum of an arithmetic series from the first to the *k*th terms, find a_1 and a_k, and then use the formula, S_k, given earlier.

EXAMPLE

Q. Find the 60th term of the arithmetic sequence: $4, 7, 10, 13, \ldots$

A. $a_{60} = 181$. The easiest way to begin this problem is to first find the formula for the *n*th term. To do so, you need a_1, which is 4. You also need the common difference, *d*, which can be found by subtracting two sequential terms, for example, $a_2 - a_1 = 7 - 4 = 3$. Plugging these into the general formula and simplifying, you get: $a_n = a_1 + (n-1)d = 4 + (n-1)\cdot 3 = 4 + 3n - 3 = 3n + 1$. Now you can find the 60th term by plugging in 60 for *n*: $a_{60} = 3(60) + 1 = 180 + 1 = 181$.

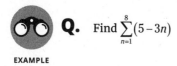

EXAMPLE

Q. Find $\sum_{n=1}^{8}(5 - 3n)$

A. –68. If you write out a few terms: $2, -1, -4, -7, \ldots$, you notice that the difference between any two consecutive terms is the same number, –3. Hence, it's an arithmetic series. To find the sum, you just have to use the arithmetic series formula. For this, you need *k* (which is 8), a_1, and a_k. You already found $a_1 = 2$. Then find $a_8 = 5 - 3(8) = 5 - 24 = -19$. Finally, plug these into the formula: $S_8 = \frac{8}{2}(2 + (-19)) = 4(-17) = -68$.

---

 7   Find the 50th term of the arithmetic sequence: $-6, -1, 4, 9, \ldots$

 8   Find the formula for the *n*th term of an arithmetic sequence where $a_1 = -3$ and $a_{15} = 53$.

9 Find the formula for the nth term of an arithmetic sequence where $a_5 = -5$ and $a_{20} = -35$.

10 Find $\sum_{n=1}^{5}\left(\frac{1}{2}n+2\right)$

11 Find $\sum_{n=4}^{10}(2n-3)$

12 Write the arithmetic series $2+\frac{7}{3}+\frac{8}{3}+3+\frac{10}{3}$ in summation notation and find the result.

Simplifying Geometric Sequences and Series

When consecutive terms in a sequence have a common ratio, the sequence is called a *geometric sequence*. To find that ratio, r, you divide any term by the term before it, and the quotient will be the same, for all pairs of terms in the sequence. Just like the other sequences, a_1 denotes the first term. To find the next term, multiply by the common ratio, r. Another pattern! The formula for the nth term of a geometric sequence is $a_n = a_1 \cdot r^{n-1}$.

TECHNICAL STUFF

As with some other sequences, you can find the sum of geometric sequences, called *geometric series*. To find a partial sum of a geometric sequence, you can use the following formula:

$S_k = \sum_{n=1}^{k} a_n = a_1\left(\frac{1-r^k}{1-r}\right)$, where $r \neq 1$.

Here, $n=1$ is the lower limit, k is the sum's upper limit, r is the common ratio, and a_1 is the first term.

You can actually find the value of an infinite sum of many geometric series. As long as r lies within the range $-1 < r < 1$, you can find the infinite sum. If $r > 1$ or $r < -1$, a_n will become arbitrarily large in absolute value, so the sum won't converge. To find the infinite sum of a geometric series where r is within the range $-1 < r < 1$, use the following formula:

$$S = \sum_{n=1}^{\infty} a_n = \frac{a_1}{1-r}$$

Geometric sequences and series are pretty easy to deal with. You just need to remember your rules for simplifying fractions. You can do it!

EXAMPLE

Q. Find the 10th term of the geometric sequence: $3, -6, 12, -24, \ldots$.

A. $a_{10} = -1{,}536$. For the formula for the nth term of a geometric sequence, you need a_1 and r. a_1 is given in the problem: 3. To find r, all you need to do is divide a_2 by a_1: $\frac{-6}{3} = -2$. Now you can simply plug these values into the formula: $a_{10} = 3 \cdot (-2)^{10-1} = 3 \cdot (-2)^9 = 3 \cdot (-512) = -1{,}536$

Q. Find the sum: $\sum_{n=1}^{5} 6\left(\frac{1}{3}\right)^{n-1}$.

A. $\frac{242}{27}$. To use the partial sum formula,

$$S_k = \sum_{n=1}^{k} a_n = a_1\left(\frac{1-r^k}{1-r}\right),$$ you need to

know a_1, r, and k. From the problem, you can identify r as $\frac{1}{3}$ and k as 5.

To find a_1, simply plug in 1 for n:

$$a_1 = 6\left(\frac{1}{3}\right)^{1-1} = 6\left(\frac{1}{3}\right)^0 = 6(1) = 6.$$ Now all

you have to do is plug the values into the formula:

$$\sum_{n=1}^{5} 6\left(\frac{1}{3}\right)^{n-1} = 6\left(\frac{1-\left(\frac{1}{3}\right)^5}{1-\frac{1}{3}}\right) = 6\left(\frac{1-\frac{1}{243}}{\frac{2}{3}}\right)$$

$$= 6\left(\frac{\frac{242}{243}}{\frac{2}{3}}\right) = 6^2 \cdot \frac{242^{121}}{_{27}243} \cdot \frac{\cancel{3}}{\cancel{2}} = \frac{242}{27}$$

13 Find the 16th term of a geometric sequence given $a_1 = 5$ and $a_2 = -15$.

14 Find the 8th term of a geometric sequence given $a_2 = 6$ and $a_6 = 486$.

15 Find the sum: $\sum_{n=1}^{6} 4\left(-\dfrac{1}{2}\right)^{n-1}$

16 Find the partial sum of the geometric series: $\dfrac{1}{6} + \dfrac{1}{3} + \dfrac{2}{3} + \cdots + \dfrac{32}{3}$.

17 Find the sum of the infinite geometric series: $\dfrac{2}{3} + \dfrac{1}{3} + \dfrac{1}{6} + \cdots$

18 Find the sum: $\sum_{n=1}^{\infty} 3\left(-\dfrac{2}{3}\right)^{n-1}$

Expanding Polynomials Using the Binomial Theorem

Binomials are polynomials with exactly two terms. Often, binomials are raised to powers to complete computations, and when you multiply out a binomial so that it doesn't have any parentheses, it's called a *binomial expansion*. One way to complete binomial expansions is to distribute terms, but if the power is high, this method can be tedious.

An easier way to expand binomials is to use the *binomial theorem*:

$$(a+b)^n = \binom{n}{0}a^n b^0 + \binom{n}{1}a^{n-1}b^1 + \binom{n}{2}a^{n-2}b^2 + \cdots + \binom{n}{n-2}a^2 b^{n-2} + \binom{n}{n-1}a^1 b^{n-1} + \binom{n}{n}a^0 b^n$$

Here, a is the first term of the binomial and b is the second term. And the coefficient for the $(r+1)$th term, $\binom{n}{r} = \dfrac{n!}{r!(n-r)!}$, is the *combinations* formula. For example, to find the binomial coefficient given by $\binom{6}{2}$, plug the values into the formula and simplify: $\binom{6}{2} = \dfrac{6!}{2!(6-2)!} = \dfrac{6!}{2!4!} = \dfrac{\overset{3}{\cancel{6}} \cdot 5 \cdot \cancel{4 \cdot 3 \cdot 2 \cdot 1}}{\cancel{2} \cdot 1 \cdot \cancel{4 \cdot 3 \cdot 2 \cdot 1}} = \dfrac{15}{1} = 15.$

Q. Write the expansion of $(3x-2)^4$.

EXAMPLE

A. $81x^4 - 216x^3 + 216x^2 - 96x + 16$. To expand, simply replace a with $3x$, b with -2, and n with 4 to get $\binom{4}{0}(3x)^4(-2)^0 + \binom{4}{1}(3x)^3(-2)^1 + \binom{4}{2}(3x)^2(-2)^2 + \binom{4}{3}(3x)^1(-2)^3 + \binom{4}{4}(3x)^0(-2)^4$.

Now, to simplify, start with the combinations formula for each term:
$(1)(3x)^4(-2)^0 + (4)(3x)^3(-2)^1 + (6)(3x)^2(-2)^2 + (4)(3x)^1(-2)^3 + (1)(3x)^0(-2)^4$.

Then, raise the monomials to the specified powers:
$(1)(81x^4)(1) + (4)(27x^3)(-2) + (6)(9x^2)(4) + (4)(3x)(-8) + (1)(1)(16)$.

Finally, multiply and simplify: $81x^4 - 216x^3 + 216x^2 - 96x + 16$.

19 Find the coefficient of $x^8 y^4$ in $(x+y)^{12}$.

 20 Find the coefficient of $x^3 y^7$ in $(2x-3y)^{10}$.

 21 Expand $(k-4)^5$.

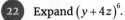 **22** Expand $(y+4z)^6$.

Answers to Problems on Sequences, Series, and Binomials

This section contains the answers for the practice problems presented in this chapter. There are also explanations available.

(1) Write the first five terms of the sequence: $a_n = \dfrac{n-3}{2n}$. The answer is $-1, -\dfrac{1}{4}, 0, \dfrac{1}{8}, \dfrac{1}{5}$.

To find each term, you plug the number of the term (n) into the formula: $a_1 = \dfrac{1-3}{2(1)} = \dfrac{-2}{2} = -1$,

$a_2 = \dfrac{2-3}{2(2)} = \dfrac{-1}{4} = -\dfrac{1}{4}$, $a_3 = \dfrac{3-3}{2(3)} = \dfrac{0}{6} = 0$, $a_4 = \dfrac{4-3}{2(4)} = \dfrac{1}{8}$, $a_5 = \dfrac{5-3}{2(5)} = \dfrac{2}{10} = \dfrac{1}{5}$.

(2) Write a general expression for the sequence to find the nth term: $2, 4, 10, 28, 82, \ldots$. The answer is $a_n = 3^{n-1} + 1$.

You're looking for a pattern where the numbers grow more rapidly as the terms progress. You see that if you subtract 1 from each term, the pattern becomes apparent: $1, 3, 9, 27, 81, \ldots$. These are powers of 3. In fact, they can be found by taking 3 to sequential powers: $3^0, 3^1, 3^2, 3^3, 3^4, \ldots$, where the power is 1 less than the number of the term, or $n-1$. So, putting that all together, you get $a_n = 3^{n-1} + 1$.

(3) Write a general expression for the sequence to find the nth term: $2, 2, \dfrac{8}{3}, 4, \dfrac{32}{5}, \ldots$. The answer is $a_n = \dfrac{2^n}{n}$.

TIP

Here's the hint for this one: The third term has a 3 in the denominator, and the fifth term has a 5 in the denominator. If you write each as an unreduced fraction over the term, n, the pattern for the denominator reveals itself: $\dfrac{2}{1}, \dfrac{4}{2}, \dfrac{8}{3}, \dfrac{16}{4}, \dfrac{32}{5}, \ldots$. The denominator is n. Now you just have to figure out the pattern for the numerator. Easy! They're all powers of 2. The first term is 2^1, the second is 2^2, and so on. So, your general expression is simply $a_n = \dfrac{2^n}{n}$.

(4) Write the next two terms of the sequence: $1, 1, 2, 3, 7, 16, \ldots$. The answer is 65, 321.

This is one of those recursive sequences. This pattern is found by adding the first term squared to the second term, and so on: $1^2 + 1 = 2$, $1^2 + 2 = 3$, $2^2 + 3 = 7$, $3^2 + 7 = 16$, $\ldots$. So, to find the next two terms, you just need to continue the pattern: $7^2 + 16 = 65$ and $16^2 + 65 = 321$.

(5) Find the sum of the first five terms of the series whose nth term is described by $2^{n-1} + 1$. The answer is 36.

Simply plug in values 1 through 5 for n and simplify: $2^{1-1} + 1 = 2$, $2^{2-1} + 1 = 3$, $2^{3-1} + 1 = 5$, $2^{4-1} + 1 = 9$, $2^{5-1} + 1 = 17$. Now you just need to add the terms: $2 + 3 + 5 + 9 + 17 = 36$.

(6) Find the sum of the first five terms of the series whose nth term is described by $3^n + 2n$. The answer is 393.

Plug in the values to find the terms and then add them up: $3^1 + 2(1) = 5$, $3^2 + 2(2) = 13$, $3^3 + 2(3) = 33$, $3^4 + 2(4) = 89$, $3^5 + 2(5) = 253$. Now just add them up: $5 + 13 + 33 + 89 + 253 = 393$.

(7) Find the 50th term of the arithmetic sequence: $-6, -1, 4, 9, \ldots$. The answer is $a_{50} = 239$.

To start, find the formula for the nth term. For this, you need a_1, which is -6. Next, you need the common difference, d, found by subtracting two sequential terms: $a_2 - a_1 = -1 - (-6) = 5$. From here, simply plug these into the general formula and simplify: $a_n = -6 + (n-1)5 = -6 + 5n - 5 = 5n - 11$. Now that you have the general formula, you can find the 50th term by plugging in 50 for n: $a_{50} = 5(50) - 11 = 250 - 11 = 239$.

(8) Find the general formula of an arithmetic sequence where $a_1 = -3$ and $a_{15} = 53$. The answer is $a_n = 4n - 7$.

Here, you have the first term, a_1, so you just need to find d. You can use a_{15} to find it. Simply plug in a_1, a_{15}, and $n = 15$ into the general formula and solve using algebra: $53 = -3 + (15-1)d$; $53 = -3 + 14d$; $56 = 14d$; $4 = d$ Now you can plug it in to find the general formula: $a_n = -3 + (n-1)4 = -3 + 4n - 4 = 4n - 7$.

(9) Find the general formula of an arithmetic sequence where $a_5 = -5$ and $a_{20} = -35$. The answer is $a_n = -2n + 5$.

To start, recognize that you have two terms, which means you can create two equations. Then you have a system of equations (check out Chapter 13 for a refresher). Solve these to find the missing variables and you can write your general formula. Start by writing your two equations with your given values and simplify: $-5 = a_1 + (5-1)d = a_1 + 4d$ and $-35 = a_1 + (20-1)d = a_1 + 19d$. Next, use elimination to solve the system by multiplying the first equation by -1 and adding the two together to get $-30 = 15d$, giving you $d = -2$. Now, simply substitute this back into either equation to find a_1: $-5 = a_1 + 4(-2)$; $-5 = a_1 - 8$; $3 = a_1$. Finally, plug in a_1 and d to find the general formula: $a_n = 3 + (n-1)(-2)$; $a_n = 3 - 2n + 2$; $a_n = -2n + 5$.

(10) Find $\sum_{n=1}^{5}\left(\frac{1}{2}n + 2\right)$. The answer is $\frac{35}{2}$.

To find the sum, you just have to use the arithmetic series formula. For this, you need k (which is 5), a_1, and a_5. Start by finding a_1: $a_1 = \frac{1}{2}(1) + 2 = \frac{5}{2}$. Then find a_5: $a_5 = \frac{1}{2}(5) + 2 = \frac{9}{2}$.

Finally, plug these into the formula: $S_5 = \left(\frac{5}{2}\right)\left(\frac{5}{2} + \frac{9}{2}\right) = \left(\frac{5}{2}\right)(7) = \frac{35}{2}$.

(11) Find $\sum_{n=4}^{10}(2n - 3)$. The answer is 77.

Follow the same steps as in Question 10. However, notice that the lower limit is 4, so you need to start by finding a_4: $2(4) - 3 = 5$, and this is like your a_1. The number of terms from 4 to 10 is 7, so $k = 7$. You also need $a_{10} = 2(10) - 3 = 17$. Finally, just plug in the values:

$S_7 = \left(\frac{7}{2}\right)(5 + 17) = \left(\frac{7}{2}\right)(22) = 77$.

(12) Write the arithmetic series $2 + \frac{7}{3} + \frac{8}{3} + 3 + \frac{10}{3}$ in summation notation and find the result. The answer is $\sum_{n=1}^{5}\left(\frac{5}{3} + \frac{1}{3}n\right) = \frac{40}{3}$.

For summation notation, you need to find the general formula and know how many terms you're dealing with. In this case, you have five terms. Therefore, you know your upper limit, k, is 5. For the general formula, you need the first term, $a_1 = 2$, and the common difference, d, which is found by subtracting two sequential terms: $a_2 - a_1 = \frac{7}{3} - 2 = \frac{1}{3} = d$. Plug these in and simplify to find your general formula: $a_n = 2 + (n-1)\frac{1}{3} = 2 + \frac{1}{3}n - \frac{1}{3} = \frac{5}{3} + \frac{1}{3}n$. Then, plug the general formula into the summation notation and add the values given in the original problem to find the result: $\sum_{n=1}^{5} \left(\frac{5}{3} + \frac{1}{3}n\right) = \frac{5}{2}\left(2 + \frac{10}{3}\right) = \frac{5}{\cancel{2}}\left(\frac{\overset{8}{\cancel{16}}}{3}\right) = \frac{40}{3}$.

(13) Find the 16th term of a geometric sequence given $a_1 = 5$ and $a_2 = -15$. The answer is $a_{16} = -71,744,535$.

To find the 16th term, you need to find the general formula. For that, you need a_1 and r. a_1 is given in the problem: 5. To find r, all you need to do is divide a_2 by a_1: $r = \frac{-15}{5} = -3$. Now you can simply plug these values into the formula: $a_{16} = 5(-3)^{16-1} = 5(-3)^{15} = -71,744,535$.

(14) Find the 8th term of a geometric sequence given $a_2 = 6$ and $a_6 = 486$. The answer is 4,374.

This time you don't have the first term, so you have to set up a system of equations and solve it: $6 = a_1 \cdot r^{2-1}$; $6 = a_1 \cdot r^1$ and $486 = a_1 \cdot r^{6-1}$; $486 = a_1 \cdot r^5$. Solve for a_1 in each and use substitution to solve for r: $a_1 = \frac{6}{r}$ and $a_1 = \frac{486}{r^5}$ gives you $\frac{6}{r} = \frac{486}{r^5}$ and $6r^5 = 486r$. You can divide both sides by r, because the ratio can't be 0 and by 6, giving you $r^4 = 81$. The ratio r can be 3 or -3. Then substitute r back into either equation to find a_1. When $r = 3$, $a_1 = 2$. And when $r = -3$, $a_1 = -2$. Now you can set up the general formulas: $a_n = 2 \cdot 3^{n-1}$ or $a_n = -2 \cdot (-3)^{n-1}$. Finally, to find the 8th term, plug in $n = 8$: $a_8 = 2 \cdot 3^7 = 4,374$ or $a_8 = -2 \cdot (-3)^7 = 4,374$. The eighth term is the same in both sequences.

(15) Find the sum: $\sum_{n=1}^{6} 4\left(-\frac{1}{2}\right)^{n-1}$. The answer is $\frac{21}{8}$.

Start by finding $a_1 = 4\left(\frac{1}{2}\right)^{1-1} = 4\left(\frac{1}{2}\right)^0 = 4$. Because $r = -\frac{1}{2}$, you have everything you need to plug into the partial sum formula:

$$\sum_{n=1}^{6} 4\left(-\frac{1}{2}\right)^{n-1} = 4\left(\frac{1 - \left(-\frac{1}{2}\right)^6}{1 - \left(-\frac{1}{2}\right)}\right) = 4\left(\frac{1 - \frac{1}{64}}{\frac{3}{2}}\right) = 4\left(\frac{\frac{63}{64}}{\frac{3}{2}}\right) = \cancel{4} \cdot \frac{\overset{21}{\cancel{63}}}{\underset{16}{\cancel{64}}} \cdot \frac{2}{\cancel{3}} = \frac{42}{16} = \frac{21}{8}$$

(16) Find the partial sum of the geometric series: $\frac{1}{6} + \frac{1}{3} + \frac{2}{3} + \cdots + \frac{32}{3}$. The answer is $\frac{127}{6}$.

Here you have a_1 and you can find r by dividing a_2 by a_1: $r = \frac{\frac{1}{3}}{\frac{1}{6}} = \frac{1}{3} \cdot \frac{6}{1} = 2$. The trick here is that you need to find n. To do so, plug the last term into the general formula and use properties of exponents to solve for n. You can now write the general formula as $a_n = \frac{1}{6} \cdot 2^{n-1}$. Solving for n in this case: $\frac{32}{3} = \frac{1}{6} \cdot 2^{n-1}$; $6 \cdot \frac{32}{3} = 6 \cdot \frac{1}{6} \cdot 2^{n-1}$; $64 = 2^{n-1}$; $2^6 = 2^{n-1}$; $6 = n-1$; $7 = n$. Now that you have all the variables, plug them in to find the partial sum: $S_7 = \frac{1}{6}\left(\frac{1 - 2^7}{1 - 2}\right) = \frac{1}{6}\left(\frac{-127}{-1}\right) = \frac{127}{6}$.

(17) Find the sum of the infinite geometric series: $\dfrac{2}{3} + \dfrac{1}{3} + \dfrac{1}{6} + \cdots$. The answer is $\dfrac{4}{3}$.

You need to start by finding r, which is a_2 divided by a_1: $\dfrac{1}{3} \div \dfrac{2}{3} = \dfrac{1}{2}$. a_1 is $\dfrac{2}{3}$. So all you need to do

is plug these values into the appropriate formula and simplify: $S = \dfrac{\dfrac{2}{3}}{1 - \dfrac{1}{2}} = \dfrac{\dfrac{2}{3}}{\dfrac{1}{2}} = \dfrac{2}{3} \cdot \dfrac{2}{1} = \dfrac{4}{3}$

(18) Find the sum: $\displaystyle\sum_{n=1}^{\infty} 3\left(-\dfrac{2}{3}\right)^{n-1}$. The answer is $\dfrac{9}{5}$.

To start, find a_1 by plugging 1 into the general formula: $3\left(-\dfrac{2}{3}\right)^{1-1} = 3\left(-\dfrac{2}{3}\right)^0 = 3(1) = 3$.

Next, notice that you have r, $-\dfrac{2}{3}$. From here, plug in these values to find the infinite sum:

$$S = \dfrac{3}{1 - \left(-\dfrac{2}{3}\right)} = \dfrac{3}{\dfrac{5}{3}} = \dfrac{3}{1} \cdot \dfrac{3}{5} = \dfrac{9}{5}$$

(19) Find the coefficient of $x^8 y^4$ in $(x+y)^{12}$. The answer is $495 x^8 y^4$.

In the expansion of the binomial, this would be the fifth term. To find the 5th term is use the

binomial theorem. In this case, $r = 5 - 1 = 4$, $a = x$, and $b = y$: $\dbinom{12}{4} x^{12-4} y^4$. Because the original

binomial doesn't have any coefficients, the coefficient just comes from the combinations for-

mula: $\dbinom{12}{4} = \dfrac{12!}{4!8!} = 495$. Multiply the other terms by this and you get $495 x^8 y^4$.

(20) Find the coefficient of $x^3 y^7$ in $(2x - 3y)^{10}$. The answer is $-2{,}099{,}520 x^3 y^7$.

In this case, $r = 8 - 1 = 7$, $a = 2x$, $b = -3y$. Plug these into the binomial theorem and simplify:

$$\dbinom{10}{7}(2x)^{10-7}(-3y)^7 = \dfrac{10!}{7!3!}(2x)^3(-3y)^7 = 120(8x^3)(-2{,}187 y^7) = -2{,}099{,}520 x^3 y^7$$

(21) Expand $(k-4)^5$. The answer is $k^5 - 20k^4 + 160k^3 - 640k^2 + 1{,}280k - 1{,}024$.

To expand, simply replace a with k, b with -4, and n with 5 to get

$$\dbinom{5}{0}k^5(-4)^0 + \dbinom{5}{1}k^4(-4)^1 + \dbinom{5}{2}k^3(-4)^2 + \dbinom{5}{3}k^2(-4)^3 + \dbinom{5}{4}k^1(-4)^4 + \dbinom{5}{5}k^0(-4)^5$$

To simplify, start with the combinations formula for each term:
$(1)k^5(-4)^0 + (5)k^4(-4)^1 + (10)k^3(-4)^2 + (10)k^2(-4)^3 + (5)k^1(-4)^4 + (1)k^0(-4)^5$.

Next, raise the monomials to the specified powers:
$(1)k^5(1) + (5)k^4(-4) + (10)k^3(16) + (10)k^2(-64) + (5)k^1(256) + (1)k^0(-1024)$.

Last, combine like terms and simplify: $k^5 - 20k^4 + 160k^3 - 640k^2 + 1{,}280k - 1{,}024$.

(22) Expand $(y+4z)^6$. The answer is $y^6 + 24y^5 z + 240y^4 z^2 + 1,280y^3 z^3 + 3,840y^2 z^4 + 6,144yz^5 + 4,096z^6$.

Here, replace a with y, b with $4z$, and n with 6 to get:

$$\binom{6}{0}y^6(4z)^0 + \binom{6}{1}y^5(4z)^1 + \binom{6}{2}y^4(4z)^2 + \binom{6}{3}y^3(4z)^3 + \binom{6}{4}y^2(4z)^4 + \binom{6}{5}y^1(4z)^5 + \binom{6}{6}y^0(4z)^6$$

Then, to simplify, start with the combinations formula for each term:

$$(1)y^6(4z)^0 + (6)y^5(4z)^1 + (15)y^4(4z)^2 + (20)y^3(4z)^3 + (15)y^2(4z)^4 + (6)y^1(4z)^5 + (1)y^0(4z)^6$$

Raise the monomials to the specified powers:

$$(1)y^6(1) + (6)y^5(4z) + (15)y^4(16z^2) + (20)y^3(64z^3) + (15)y^2(256z^4) + (6)y^1(1,024z^5) + (1)y^0(4,096z^6)$$

Finally, combine like terms and simplify:

$$y^6 + 24y^5 z + 240y^4 z^2 + 1,280y^3 z^3 + 3,840y^2 z^4 + 6,144yz^5 + 4,096z^6.$$

Chapter **15**

Previewing Calculus

The study of mathematics usually begins with Algebra I. Some of the information learned there is repeated in Algebra II and presented again in pre-calculus (each time with some more challenging aspects than found in the previous courses). The end of pre-calculus is the beginning of calculus.

Calculus is the study of change. Before calculus, much of the material used to solve problems is constant, not changing over time. For example, up until calculus, in a distance problem, the rate at which a car is moving remains constant. The slope of a straight line is always a constant. The volume of a shape is always a constant. But in calculus, all these can move and grow and change. For example, the car can accelerate, decelerate, and accelerate again, all within the same problem, which changes the whole outcome. The line can now be a curve so that its slope changes over time. The shape that you're trying to find the volume of can get bigger or smaller, so that the volume changes over time.

All this change opens up so many possibilities to you. And this chapter is designed to help you make the leap.

Finding Limits: Graphically, Analytically, and Algebraically

Graphing functions can be simple, or, as you expand your horizons, it can become more complex and intricate. The more complicated the function is, the more complicated the graph tends to be. You've seen functions that are undefined at certain values; the graph has either a hole or a vertical asymptote, which affects your domain. In this section, you look at the *limit* of a function at a point — what the function would do if it could.

REMEMBER

In symbols, a limit is written as $\lim_{x \to n} f(x) = L$, which is read as "the limit of $f(x)$ as x approaches n is L." L is the output value that you're looking for. You're looking at the function $f(x)$ as the input value x approaches some chosen value n. L is the limit if the values of $f(x)$ can be made as close as you need to L by taking x sufficiently close to (but not equal to) n. For the limit of a function to exist, the left limit and the right limit must be equal.

> » A left limit of a given function $f(x)$ as x approaches n is a value that $f(x)$ tends to no matter how x approaches n from the left side of n. This is written as $\lim_{x \to n^-} f(x) = L$.
>
> » A right limit of a given function $f(x)$ as x approaches n is a value that $f(x)$ tends to no matter how x approaches n from the right side of n. This is written as $\lim_{x \to n^+} f(x) = L$.

Only when the left limit and right limit are the same does the function have a limit. When $\lim_{x \to n^-} f(x) = \lim_{x \to n^+} f(x) = L$, then $\lim_{x \to n} f(x) = L$.

You can find a limit in three different ways: graphically, analytically, and algebraically. In the following sections, you see each method so you know which to choose.

TIP

Before you try any of the techniques for finding a limit, always try plugging in the value that x is approaching into the function. It's also recommended using the graphing method when you've been given the graph and asked to find the limit. The analytical method works for most functions, and sometimes it's the only method that will work. However, if the algebraic method works, then go with it — the analytical method can take a bit longer.

Graphically

When you're given the graph of a function and asked to find the limit, just read the graph as x approaches the given value and see what the y value would have been (or was if the function is defined).

Q. In the given graph for $f(x)$, find $\lim_{x \to -3} f(x)$, $\lim_{x \to 4} f(x)$, and $\lim_{x \to 6} f(x)$.

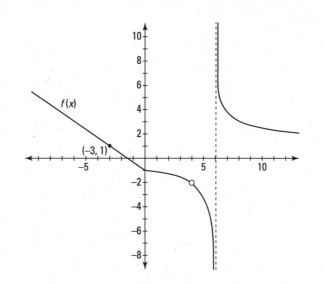

A. $\lim_{x \to -3} f(x) = 1$, $\lim_{x \to 4} f(x) = -2$, and $\lim_{x \to 6} f(x)$ doesn't exist. The function is defined at $x = -3$ because you can see a point there. The limit of the function as x tends to -3 exists and is 1. The y-values are near 1 when x is close to -3, so $\lim_{x \to -3} f(x) = 1$.

The function isn't defined at $x = 4$ because the graph has a hole, but if you move along the graph from the left as x approaches 4, y also approaches a value of -2. Because this is the same from the right side, $\lim_{x \to 4} f(x) = -2$. The graph has a vertical asymptote at $x = 6$, and you can see that the function values are increasingly positive as x approaches 6 from the right and increasingly negative as x approaches 6 from the left. Thus, the limit as x approaches 6 doesn't exist. You can indicate this with *DNE* (does not exist).

---

1 In the given graph for $g(x)$, find $\lim_{x \to -5} g(x)$, $\lim_{x \to -2} g(x)$, and $\lim_{x \to 1} g(x)$.

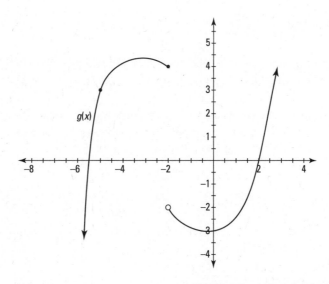

2 In the given graph for $h(x)$, find $\lim\limits_{x \to -3} h(x)$, $\lim\limits_{x \to 5} h(x)$, and $\lim\limits_{x \to 0} h(x)$.

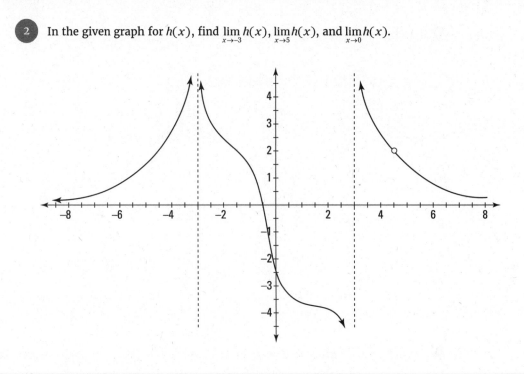

Analytically

Analytically means *systematically*, and that's exactly how you find the limit using this method. This method is also called *numerically* estimating a limit. It should be pointed out that this method is not foolproof. You set up a chart, and the value that x is approaching goes in the middle (in the middle of the top row, if you're working horizontally). Before the value you're seeking, you put values smaller than x that get closer to the value that x is approaching, and you do the same thing after the x picking larger values. The second row or column should be the y values when you plug the x values into the function. Observe what you find, and, hopefully, you have your limit!

Q. Find $\lim\limits_{x \to 2} \dfrac{x^2 + 2x - 8}{x - 2}$.

EXAMPLE

A. $\lim\limits_{x \to 2} \dfrac{x^2 + 2x - 8}{x - 2} = 6$. Notice that if you plug 2 into the function for x, you get 0 in the denominator. That means the function is undefined there. If you're asked to find this limit analytically, set up a chart. Here's a table that has been set up. *Note:* Your chart doesn't have to look exactly like this; no one way works all the time in finding the limit analytically. Just try to get those x values really close to the value you're approaching — that's usually the best way to find the limit.

x	1	1.9	1.99	1.999	2	2.001	2.01	2.1	3
y	5	5.9	5.99	5.999	$\lim\limits_{x \to 2}$	6.001	6.01	6.1	7

Notice that the y values from both the left and the right seem to be approaching 6; that's your limit.

3 Find $\lim\limits_{x \to -1} \dfrac{x^2 + 13x + 12}{x + 1}$ numerically.

4 Find $\lim\limits_{x \to 2} \dfrac{3x^2 - 5x - 2}{x - 2}$ numerically.

Algebraically

To find a limit algebraically, you have available to you four different techniques: plugging in, factoring, rationalizing the numerator or denominator, and finding the lowest common denominator. Always start by plugging the given number into the function just to see whether it works. If the answer is undefined, then move on to one of the other three techniques — each one depending on the given function.

Plugging in

This first technique asks you to substitute the given value into the function. If you get an undefined value, like 0 in the denominator of the fraction, try something else, unless the numerator approaches something nonzero, in which case, you automatically know that the limit doesn't exist. But when substitution works, this technique is the quickest way to find a limit. But remember, the value has to be the same coming from both sides of the number, or the limit doesn't exist.

Factoring

When the function is a rational function with polynomials in the numerator and the denominator, try factoring them. If you need a refresher on factoring, see Chapter 4. Chances are some factors will cancel from both the numerator and the denominator. You can then substitute the given value into the cancelled version and often get an answer that's also your limit.

If you still get an undefined function when you follow the steps of factoring the rational function on the top and bottom, canceling, and plugging in the given value, then the limit does not exist (DNE).

Rationalizing the numerator

When you see square roots in the rational function in the numerator and plugging in doesn't work, always try to rationalize the numerator or denominator. That's right — you multiply by the conjugate of the radical expression on both the top and bottom of the fraction. When you do, you usually see several terms cancel, and the function simplifies down to a point where you can plug in the given value and find the limit.

Finding the lowest common denominator

When a rational function involves more than one rational term, find the common denominators and add or subtract terms, then cancel and simplify. You can then plug in the given value to find the limit.

Q. Find $\lim\limits_{x \to 0} \dfrac{\frac{1}{x+5} - \frac{1}{5}}{x}$

A. $\lim\limits_{x \to 0} \dfrac{\frac{1}{x+5} - \frac{1}{5}}{x} = -\dfrac{1}{25}$. Find the common denominator of the fractions on the top first:

$\lim\limits_{x \to 0} \dfrac{\frac{1}{x+5} - \frac{1}{5}}{x} = \lim\limits_{x \to 0} \dfrac{\frac{1}{x+5} \cdot \frac{5}{5} - \frac{1}{5} \cdot \frac{x+5}{x+5}}{x} = \lim\limits_{x \to 0} \dfrac{\frac{5}{5(x+5)} - \frac{x+5}{5(x+5)}}{x}$. Subtract and then simplify

the top now that you have a common denominator: $= \lim\limits_{x \to 0} \dfrac{\frac{5-(x+5)}{5(x+5)}}{x} = \lim\limits_{x \to 0} \dfrac{\frac{-x}{5(x+5)}}{x} =$

$\lim\limits_{x \to 0} \dfrac{-x}{5(x+5)} \cdot \dfrac{1}{x} = \lim\limits_{x \to 0} \dfrac{-\cancel{x}}{5(x+5)} \cdot \dfrac{1}{\cancel{x}} = \lim\limits_{x \to 0} \dfrac{-1}{5(x+5)}$. Notice that you can now plug 0 into the

last expression and get the limit: $= \dfrac{-1}{5(0+5)} = \dfrac{-1}{25}$.

---

5 Find $\lim\limits_{x \to 2} \dfrac{3x^2 - 5x - 2}{x - 2}$ algebraically.

6 Find $\lim\limits_{x \to 5} \dfrac{\sqrt{x-1} - 2}{x - 5}$

Knowing Your Limits

Mathematics provides you with a few limit laws that can help you find the limits of combined functions: adding, subtracting, multipling, dividing, and even raising to powers. If you can find the limit of each individual function, you can find the limit of the combined function as well.

TECHNICAL
STUFF

If $\lim_{x \to n} f(x) = L$ and $\lim_{x \to n} g(x) = M$, then:

>> **Addition law:** $\lim_{x \to n}(f(x) + g(x)) = L + M$

>> **Subtraction law:** $\lim_{x \to n}(f(x) - g(x)) = L - M$

>> **Multiplication law:** $\lim_{x \to n}(f(x) \cdot g(x)) = L \cdot M$

>> **Division law:** $\lim_{x \to n}\left(\dfrac{f(x)}{g(x)}\right) = \dfrac{L}{M}$, provided $M \neq 0$

>> **Power law:** $\lim_{x \to n}(f(x))^p = L^p$

For this whole section, use the following to answer the questions:

$$\lim_{x \to 1} f(x) = -5 \qquad \lim_{x \to 1} g(x) = 2 \qquad \lim_{x \to 1} h(x) = 0$$

Q. Find $\lim_{x \to 1}(g(x) - f(x))$.

A. $\lim_{x \to 1}(g(x) - f(x)) = 7$. The limit of $g(x)$ is 2 and the limit of $f(x)$ is -5. To find $\lim_{x \to 1}(g(x) - f(x))$, use the subtraction law: $2 - (-5) = 7$. It really is that easy!

---

7 Find $\lim_{x \to 1} \dfrac{f(x)}{g(x)}$

8 Find $\lim_{x \to 1}\left((f(x))^2 - 2h(x) + \dfrac{1}{g(x)}\right)$

9 Find $\lim\limits_{x \to 1} \dfrac{\sqrt{g(x)} - f(x)}{5g(x)}$

 10 Find $\lim\limits_{x \to 1} \dfrac{f(x)}{h(x)}$

Calculating the Average Rate of Change

The *average rate of change* of a line is simply its slope. But the average rate of change of a curve is an ever-changing number, depending on where you're looking on the curve.

Consider the graph of $f(x) = x^2$, the basic parabola. If you draw tangents to the curve at various points, some of the lines will have positive slopes, some negative, and one of the lines will have a slope of 0, at the vertex of the parabola.

A *tangent to a curve* at a point is a straight line that touches the curve at just that one point.

REMEMBER Concentrating on just one tangent line at one point, consider the tangent to the curve at the point $(2, 4)$. The tangent drawn to the point $(2, 4)$ touches at just that single point. But how can you find the slope (average rate of change) of that tangent line when you have just one point? The slope formula requires two points.

One method you can use is to pick two points on the curve, one on each side of the point in question. In this case, consider the points $(1.5, 2.25)$ and $(2.5, 6.25)$. When a line is drawn through the points, you have it intersecting the parabola.

You can compute the slope of the line through the two points and use that as an estimate of the slope of the tangent line through $(2, 4)$.

$$m = \frac{6.25 - 2.25}{2.5 - 1.5} = \frac{4}{1} = 4$$

You can then try two points a little closer to the point on the curve, like $(1.9, 3.61)$ and $(2.1, 4.41)$. Compute the slope of the line through those points. And this can continue choosing points closer and closer to the target, $(2, 4)$. (Does this sound familiar — like a limit?)

This process can give you a good estimate of the average rate of change at a point. But the even better news is that calculus will provide a much nicer and easier method for finding this value — the method is called finding the derivative.

11 Estimate the slope of the tangent to the cubic $y = x^3 - 3x + 1$ at the point $(1, -1)$ using $x = 0$ and $x = 2$.

12 Estimate the slope of the tangent to the cubic $y = x^3 - 3x + 1$ at the point $(1, -1)$ using $x = 0.9$ and $x = 1.1$.

Determining Continuity

The word "continuity" means the same thing in math as it does in your everyday life. Something that's continuous has a stability or a permanence to it . . . it never stops. In pre-calculus, you see functions that have holes in their graph, jumps in their graph, or asymptotes — just to name a few discontinuities. A graph that doesn't have holes, jumps, or vertical asymptotes keeps going forever, and we call that function *continuous*.

TIP

When investigating continuity, you usually look at specific values in the domain to determine continuity instead of looking at the entire function. Even discontinuous functions are discontinuous only at certain places. The discontinuity at a certain x value in any function is termed either *removable* (a hole in the graph) or *nonremovable*. In the case of rational functions that have discontinuities due to a factor in the denominator that goes to zero at the value c, it all depends on the factored versions of the polynomials in the numerator and denominator. If there is an instance of the factor causing the discontinuity to cancel out of the denominator when reducing the fraction, then the discontinuity at c is removable. If not, the discontinuity is nonremovable.

REMEMBER

Three things must be true for a function to be continuous at $x = c$:

>> $f(c)$ **must be defined.** When you plug c into the function, you must get a value out either by definition or through a specific rule. For example, getting 0 in the denominator is unacceptable and therefore a discontinuity.

>> **The limit of the function as x approaches c must exist.** The left and right limits must be the same. If they aren't, the function is discontinuous there.

>> **The function's value and the limit must be the same.** If the value of the function is one thing and the limit is something different, that's not good; the function is discontinuous there.

Here's the graph of a function where each one of the preceding situations fails:

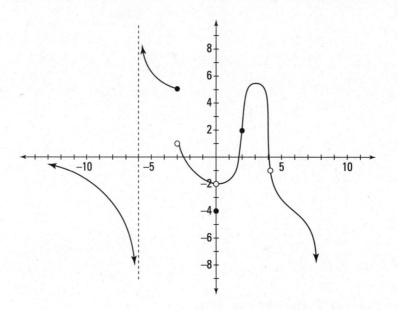

At $x = 4.3$, the graph has a hole. The function is undefined there, and therefore discontinuous at $x = 4.3$.

At $x = -3$ the function jumps. The limit as x approaches -3 from the left is 5, and from the right the limit is 1, so the limit doesn't exist, and the function is discontinuous.

At $x = 0$ the function is defined at one point: $f(0) = -4$, but the limit as x approaches 0 from the left and from the right is -2. These two values must be the same for the function to be continuous.

One point where the function is continuous is at $x = 2$: $\lim\limits_{x \to 2} f(x) = 2 = f(2)$.

Q. Is $f(x) = \dfrac{3}{x+2}$ continuous at $x = 3$?

A. $f(x)$ is continuous at $x = 3$. This is a rational function and $x = 3$ is in its domain.

Q. Explain why $f(x) = \dfrac{3}{x+2}$ is not continuous at $x = -2$. Is this discontinuity removable or nonremovable?

A. The function is not defined at $x = -2$. The value $x = -2$ creates a 0 in the denominator, and this factor cannot be removed by factoring. Therefore the discontinuity that exists is nonremovable.

13 Determine whether $g(x) = \dfrac{x^2 - 4x - 5}{x^2 - 2x - 15}$ is continuous at $x = 5$.

14 Is $g(x) = \dfrac{x^2 - 4x - 5}{x^2 - 2x - 15}$ continuous at $x = 0$?

15 Is $g(x) = \dfrac{x^2 - 4x - 5}{x^2 - 2x - 15}$ continuous at $x = -3$?

16 Determine whether $h(x) = \begin{cases} 3x - 1 & \text{if } x \le -2 \\ x^2 - 11 & \text{if } x > -2 \end{cases}$ is continuous at $x = -2$.

17 Determine whether $p(x) = \begin{cases} \dfrac{1}{2}x - 3 & \text{if } x \le 1 \\ 4x + 3 & \text{if } x > 1 \end{cases}$ is continuous at $x = 1$.

18 Determine if $q(x) = \begin{cases} \cos x + 1 & \text{if } x < 0 \\ e^x + 1 & \text{if } x \ge 0 \end{cases}$ is continuous at $x = 0$.

Answers to Problems on Calculus

Following are the answers to problems dealing with calculus. You also find guidance on getting the answers.

1 In the given graph for $g(x)$, find $\lim\limits_{x\to-5} g(x)$, $\lim\limits_{x\to-2} g(x)$, and $\lim\limits_{x\to1} g(x)$. The answers are $\lim\limits_{x\to-5} g(x) = 3$, $\lim\limits_{x\to-2} g(x)$ *DNE* and $\lim\limits_{x\to1} g(x) \approx -2.6$.

Looking at the graph for $g(x)$, when x tends to -5, $g(x)$ tends to 3. When x approaches -2 from the left, the limit is 4, but when x approaches -2 from the right, the limit is -2. Because these two values aren't the same, $\lim\limits_{x\to-2} g(x)$ does not exist (*DNE*). And when x approaches 1, the graph has a y value somewhere between -2 and -3. Because you can't be exact using a graph, just do your best to approximate the limit. It looks like it's about -2.6.

2 In the given graph for $h(x)$, find $\lim\limits_{x\to-3} h(x)$, $\lim\limits_{x\to5} h(x)$, and $\lim\limits_{x\to0} h(x)$. The answers are $\lim\limits_{x\to-3} h(x) = \infty$, $\lim\limits_{x\to5} h(x) = 2$, and $\lim\limits_{x\to0} h(x) \approx -2.5$.

This answer also comes directly from the given graph. Trace your fingers along the graph as x gets closer and closer to -3 from the left and notice that the line keeps going up. Also notice that as x approaches this same value from the right, the graph is also going up — forever and ever up. That's why $\lim\limits_{x\to-3} h(x)$ DNE (does not exist). Meanwhile, as x approaches 5, $h(x)$ approaches 2, and as x approaches 0, it looks like $h(x)$ is very close to -2.5.

3 Find $\lim\limits_{x\to-1} \dfrac{x^2 + 13x + 12}{x+1}$ numerically. The answer is 11.

See the following chart for the analytical evaluation of this limit.

x	-2	-1.1	-1.01	-1.001	-1	-0.999	-0.99	-0.9	0
y	10	10.9	10.99	10.999	$\lim\limits_{x\to-1}$	11.001	11.01	11.1	12

By looking at the y values in the second row, it looks like from both the left and the right y is approaching 11.

4 Find $\lim\limits_{x\to2} \dfrac{3x^2 - 5x - 2}{x-2}$ numerically. The answer is 7.

Here's a chart used for this limit.

x	1	1.9	1.99	1.999	2	2.001	2.01	2.1	3
y	4	6.7	6.97	6.997	$\lim\limits_{x\to2}$	7.003	7.03	7.3	10

5 Find $\lim\limits_{x\to2} \dfrac{3x^2 - 5x - 2}{x-2}$ algebraically. The answer is 7.

Plugging 2 into the function gives you a 0 in the denominator and also in the numerator, so you must try another technique. This rational function has a numerator that factors. (You found the limit numerically in Question 4.) When you factor it, you get $\lim\limits_{x\to2} \dfrac{(3x+1)(x-2)}{x-2}$.

This reduces to $\lim\limits_{x\to2}(3x+1)$, which gives you a function that you *can* plug 2 into: $3(2) + 1 = 7$.

6 Find $\lim\limits_{x \to 5} \dfrac{\sqrt{x-1}-2}{x-5}$. The answer is $\dfrac{1}{4}$.

Substitute 5 into this expression and you get 0 in both the denominator and in the numerator. Noticing that the numerator has a square root, you multiply by the conjugate of the numerator to rationalize the numerator.

Multiplying by the conjugate: $\lim\limits_{x \to 5} \dfrac{\sqrt{x-1}-2}{x-5} \cdot \dfrac{\sqrt{x-1}+2}{\sqrt{x-1}+2}$. Multiply and simplify in the numerator.

However, don't multiply out the denominators — the final expression will reduce more easily if you don't:

$$= \lim_{x \to 5} \frac{x-1-4}{(x-5)\left(\sqrt{x-1}+2\right)} = \lim_{x \to 5} \frac{\cancel{x-5}}{\cancel{(x-5)}\left(\sqrt{x-1}+2\right)} = \lim_{x \to 5} \frac{1}{\sqrt{x-1}+2}.$$

Now when you plug in 5, $\lim\limits_{x \to 5} \dfrac{1}{\sqrt{x-1}+2} = \dfrac{1}{\sqrt{5-1}+2} = \dfrac{1}{\sqrt{4}+2} = \dfrac{1}{4}$.

For problems 7 – 10, use the following: $\lim\limits_{x \to 1} f(x) = -5$ $\lim\limits_{x \to 1} g(x) = 2$ $\lim\limits_{x \to 1} h(x) = 0$.

7 Find $\lim\limits_{x \to 1} \dfrac{f(x)}{g(x)}$. The answer is $-\dfrac{5}{2}$.

Because you know both limits, to find the limit of their quotient, divide the values:

$$\lim_{x \to 1} \frac{f(x)}{g(x)} = \frac{-5}{2}.$$

8 Find $\lim\limits_{x \to 1} \left(\left(f(x)\right)^2 - 2h(x) + \dfrac{1}{g(x)} \right)$. The answer is $\dfrac{43}{2}$.

Plug in the information that you know based on the given limits:

$$\lim_{x \to 1} \left(\left(f(x)\right)^2 - 2h(x) + \frac{1}{g(x)} \right) = (-5)^2 - 2(2) + \frac{1}{2} = 25 - 4 + \frac{1}{2} = 21\frac{1}{2} = \frac{43}{2}$$

9 Find $\lim\limits_{x \to 1} \dfrac{\sqrt{g(x)} - f(x)}{5g(x)}$. The answer is $\dfrac{\sqrt{2}+5}{10}$.

Plug in the given values: $\lim\limits_{x \to 1} \dfrac{\sqrt{g(x)} - f(x)}{5g(x)} = \dfrac{\sqrt{2} - (-5)}{5(2)} = \dfrac{\sqrt{2}+5}{10}$

10 Find $\lim\limits_{x \to 1} \dfrac{f(x)}{h(x)}$. The answer is DNE.

This time, putting the limit of $h(x)$ in the denominator also puts 0 in the denominator. The limit does not exist because the denominator approaches 0 and the numerator doesn't as x tends to 1.

$$\lim_{x \to 1} \frac{f(x)}{h(x)} = \frac{-5}{0}$$

11 Estimate the slope of the tangent to the cubic $y = x^3 - 3x + 1$ at the point $(1,-1)$ using $x = 0$ and $x = 2$. The answer is 1.

When $x = 0$, then $y = 1$, and when $x = 2$, $y = 3$. Substituting into the slope formula, you have

$$m = \frac{3-1}{2-0} = \frac{2}{2} = 1.$$

12 Estimate the slope of the tangent to the cubic $y = x^3 - 3x + 1$ at the point $(1,-1)$ using $x = 0.9$ and $x = 1.1$. The answer is 1.

When $x = 0.9$, then $y = -0.971$, and when $x = 1.1$, $y = -0.969$. Substituting into the slope formula, you have $m = \dfrac{0.969 - 0.971}{1.1 - 0.9} = \dfrac{-0.002}{0.2} = 0.1$.

(13) Determine whether $g(x) = \dfrac{x^2 - 4x - 5}{x^2 - 2x - 15}$ is continuous at $x = 5$. The answer is no, the function is not continuous at $x = 5$.

Factor the given equation first: $g(x) = \dfrac{(x-5)(x+1)}{(x-5)(x+3)}$. Reduce to get $= \dfrac{\cancel{(x-5)}(x+1)}{\cancel{(x-5)}(x+3)} = \dfrac{x+1}{x+3}$.

Notice that when you plug 5 into this simplified expression, you do get an answer of $\dfrac{6}{8}$, or $\dfrac{3}{4}$.

But this isn't the original, given equation. The graph is going to look and act like $\dfrac{x+1}{x+3}$, but because the original function is $g(x) = \dfrac{(x-5)(x+1)}{(x-5)(x+3)}$, there's still going to be a hole in the graph (try plugging 5 into either of them and see what happens). This is why g has a removable discontinuity at $x = 5$.

(14) Is $g(x) = \dfrac{x^2 - 4x - 5}{x^2 - 2x - 15}$ continuous at $x = 0$? The answer is yes, the function $g(x)$ is continuous at $x = 0$.

Observe that this is a rational function with 0 in its domain. You can simply plug 0 into this function and get: $g(0) = \dfrac{0^2 - 4(0) - 5}{0^2 - 2(0) - 15} = \dfrac{-5}{-15} = \dfrac{1}{3}$.

(15) Is $g(x) = \dfrac{x^2 - 4x - 5}{x^2 - 2x - 15}$ continuous at $x = -3$? The answer is no, the function is not continuous at $x = -3$. In fact, $x = -3$ is a nonremovable discontinuity. Plugging -3 into the original function $g(x) = \dfrac{x^2 - 4x - 5}{x^2 - 2x - 15}$ gives you 0 in the denominator, so you know right away that it's discontinuous. When you factor and simplify to $g(x) = \dfrac{x+1}{x+3}$, you *still* get 0 in the denominator, so the discontinuity is nonremovable.

(16) Determine whether $h(x) = \begin{cases} 3x - 1 & \text{if } x \le -2 \\ x^2 - 11 & \text{if } x > -2 \end{cases}$ is continuous at $x = -2$. The function is continuous at $x = -2$.

If you don't know how to deal with piecewise functions like this, get a refresher from Chapter 3.

First, look at $h(-2) = 3(-2) - 1 = -7$. The function exists at $x = -2$.

Now, look at both $\lim\limits_{x \to -2^-} h(x)$ and $\lim\limits_{x \to -2^+} h(x)$. $\lim\limits_{x \to -2^-} h(x) = 3(-2) - 1 = -7$ and $\lim\limits_{x \to -2^+} h(x) = (-2)^2 - 11 = -7$. Because the left limit matches the right limit, the function has a limit as x approaches -2. Lastly, because the function value matches the limit value, the function is continuous at $x = -2$.

(17) Determine whether $p(x) = \begin{cases} \dfrac{1}{2}x - 3 & \text{if } x \le 1 \\ 4x + 3 & \text{if } x > 1 \end{cases}$ is continuous at $x = 1$. The function isn't continuous at $x = 1$.

$p(1) = \dfrac{1}{2}(1) - 3 = -\dfrac{5}{2}$. The function exists at $x = 1$.

$\lim\limits_{x \to 1^-} p(x) = -\dfrac{5}{2}$, but $\lim\limits_{x \to 1^+} p(x) = 4(1) + 3 = 7$. These two values aren't equal, so there is no limit and the function is discontinuous at $x = 1$.

(18) Determine if $q(x) = \begin{cases} \cos x + 1 & \text{if } x < 0 \\ e^x + 1 & \text{if } x \ge 0 \end{cases}$ is continuous when $x = 0$. The function is continuous at $x = 0$. First, $q(0) = e^0 + 1 = 1 + 1 = 2$. Then, looking at the trig function,

$\lim\limits_{x \to 0}(\cos x + 1) = \cos 0 + 1 = 1 + 1 = 2$. The function is continuous when $x = 0$.

5

The Part of Tens

IN THIS PART . . .

This part has a summary of the parent graphs that are covered in Chapter 3, including how to transform them. Think of it as a quick guide to all the topics you've seen regarding graphing and transforming parent functions. This part also includes a chapter on the mistakes that are often made in pre-calculus and how to avoid them (please avoid them!).

Chapter **16**

Ten Plus Parent Graphs

A picture is worth a thousand words, and graphing is just math in pictures! These pictures can give you important information about the characteristics of a function. The most basic graphs are called parent graphs. These graphs are in their original, unshifted, unaltered form. Any parent graph can be stretched, shrunk, shifted, flipped, or a combination of these actions. Parent graphs are extremely useful because you can use them to graph a more complicated version of the same function using transformations (see Chapter 3). That way, if you're given a complex function, you automatically have a basic idea of what the graph will look like without having to plug in a whole bunch of numbers first. Essentially, by knowing what the parent looks like, you get a good idea about the kids — the apple doesn't fall far from the tree, right? In this chapter, you have a full album those family pictures!

Squaring Up with Quadratics

The basic quadratic is simplicity itself: $y = x^2$. Its graph is a parabola with a vertex at the origin, reflected over the y-axis (see Figure 16-1). You can find out more about graphing quadratics in Chapters 3 and 12.

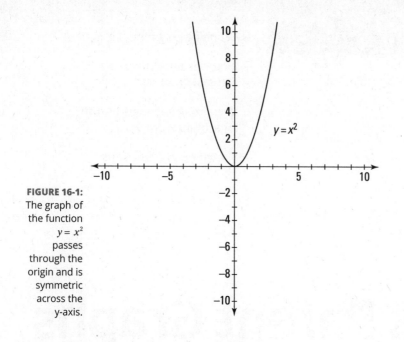

FIGURE 16-1:
The graph of
the function
$y = x^2$
passes
through the
origin and is
symmetric
across the
y-axis.

Cueing Up for Cubics

The parent graph of the cubic function, $y = x^3$, also passes through the origin. This graph is symmetric over the origin (see Figure 16-2). You find cubic functions in Chapter 3.

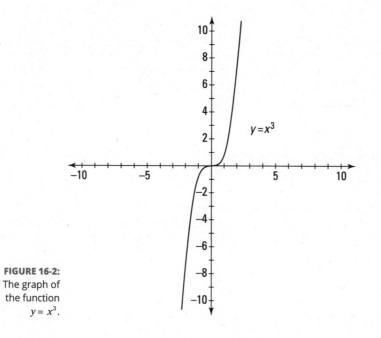

FIGURE 16-2:
The graph of
the function
$y = x^3$.

Rooting for Square Roots and Cube Roots

A square root function graph looks like a parabola that has been rotated clockwise 90 degrees and cut in half. It's cut in half (only positive) because you can't take the square root of a negative number. The parent graph is pictured in Figure 16-3.

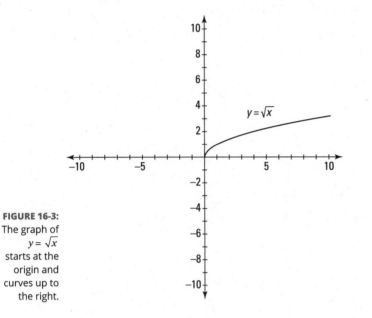

FIGURE 16-3:
The graph of
$y = \sqrt{x}$
starts at the
origin and
curves up to
the right.

Cube root functions are the inverse of cubic functions, so their graphs reflect that. The parent graph of a cube root function passes through the origin and is symmetric over it, as shown in Figure 16-4.

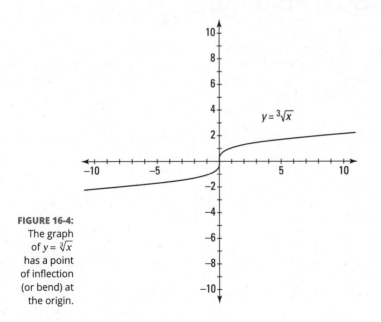

FIGURE 16-4:
The graph
of $y = \sqrt[3]{x}$
has a point
of inflection
(or bend) at
the origin.

Graphing Absolutely Fabulous Absolute Value Functions

Because the absolute value function turns all input into nonnegative values (0 or positive), the parent graph is always above the *x*-axis, except where it touches the origin. Figure 16-5 shows the parent graph in its characteristic V shape.

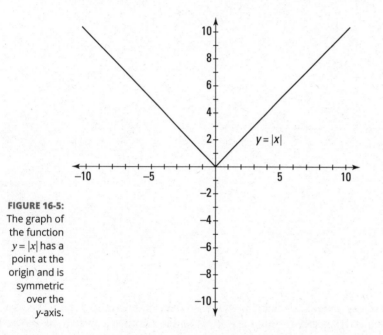

FIGURE 16-5:
The graph of the function $y = |x|$ has a point at the origin and is symmetric over the *y*-axis.

Flipping over Rational Functions

In Chapter 3, you see the steps for graphing rational functions. These involve finding asymptotes, intercepts, and key points. These functions don't really have parent graphs per se, but the most basic rational function has asymptotes involving zeros: $y = 0$ and $x = 0$ (see Figure 16-6). To see more, flip back to Chapter 3.

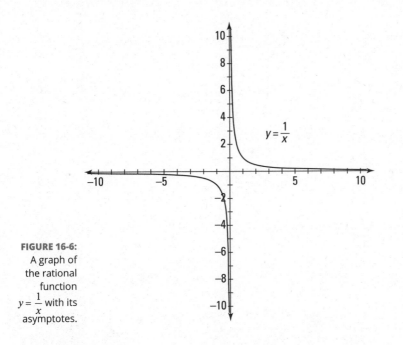

$$y = \frac{1}{x}$$

FIGURE 16-6:
A graph of the rational function $y = \frac{1}{x}$ with its asymptotes.

Exploring Exponential Graphs and Logarithmic Graphs

The parent graph of an exponential function is $y = b^x$ where b is the base. The value of b has to be some positive number. A nice example and frequently found function is the graph of $y = e^x$, where e is about 2.718. This graph passes through the point (0, 1) and has a horizontal asymptote of the x-axis, as shown in Figure 16-7. You'll find more on exponential graphs in Chapter 5.

The inverse (see Chapter 3) of an exponential function is a logarithmic function. So here you see the inverse of the graph of $y = e^x$, which is the graph of $y = \log_e x$. The log base e rates a special notation, also known as the natural log, or $y = \ln x$ (see Chapter 5). This graph passes through the point (1, 0) and has a vertical asymptote of the y-axis, as shown in Figure 16-8. Logarithmic graphs are covered in Chapter 5. Observe that exponential functions with $b > 1$ increase very rapidly, while $y = \ln x$ increases very slowly. However, they all increase without bound.

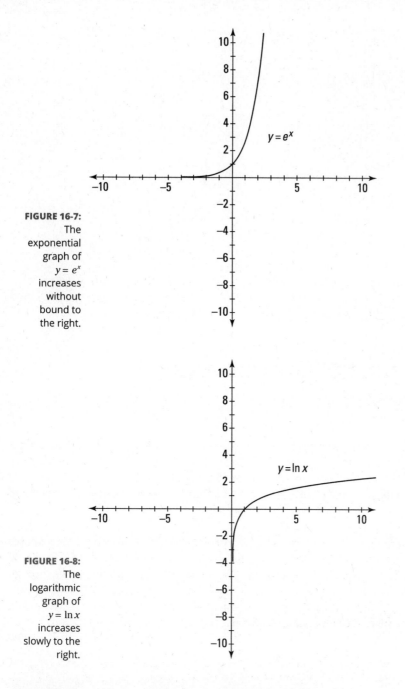

FIGURE 16-7:
The exponential graph of $y = e^x$ increases without bound to the right.

FIGURE 16-8:
The logarithmic graph of $y = \ln x$ increases slowly to the right.

Seeing the Sine and Cosine

A sine graph looks like a wave. The parent graph passes through the origin and has an amplitude of 1. The period is 2π, which means that the wave repeats itself every 2π. Figure 16-9 shows one full period of the parent sine graph between $-\pi$ and π.

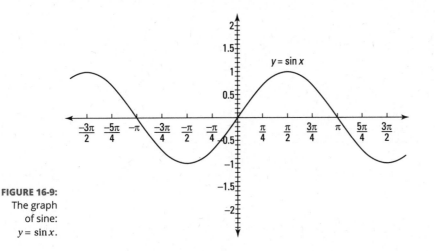

FIGURE 16-9:
The graph
of sine:
$y = \sin x$.

Like sine, the graph of cosine is a wave. This parent graph passes through the point $(0,1)$ and also has an amplitude of 1 and a period of 2π. You can see the parent graph in Figure 16-10. For more information about graphing sine and cosine, turn to Chapter 7.

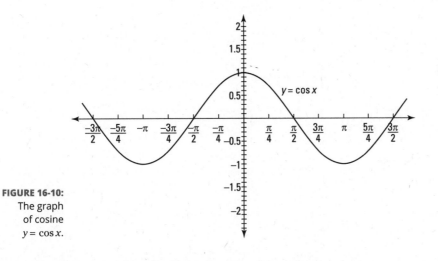

FIGURE 16-10:
The graph
of cosine
$y = \cos x$.

Covering Cosecant and Secant

Remember reciprocals? Well, cosecant is the reciprocal of sine, so the graph of cosecant reflects that. The parent sine graph and the parent cosecant graph are both depicted in Figure 16-11 so you can see the relationship. You can find specific graphing information in Chapter 7.

Again, like cosecant, secant is the reciprocal of cosine, so the graph of secant is related to the graph of cosine. To picture this, you see a lightly drawn graph of cosine along with the parent graph of secant in Figure 16-12.

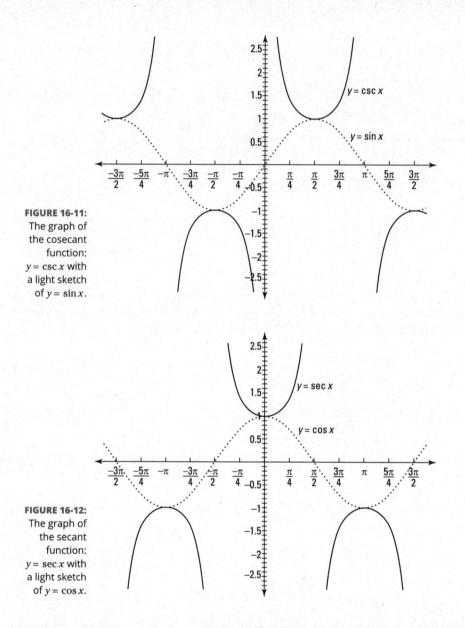

FIGURE 16-11:
The graph of
the cosecant
function:
$y = \csc x$ with
a light sketch
of $y = \sin x$.

FIGURE 16-12:
The graph of
the secant
function:
$y = \sec x$ with
a light sketch
of $y = \cos x$.

Tripping over Tangent and Cotangent

One repeating pattern of the graph of tangent is its asymptotes, where the function is undefined. Like other trig graphs, a tangent graph has a period where it repeats itself. In this case, it's π. In Figure 16-13, one period of the parent tangent graph lies between $-\frac{\pi}{2}$ and $\frac{\pi}{2}$. For more information about graphing tangents, turn to Chapter 7.

Like tangents, the parent graph of cotangent has asymptotes at regular intervals. Also like tangents, the period of cotangent is π. In Figure 16-14, you see one period of the parent cotangent graph between 0 and π. You can get more information about graphing cotangents in Chapter 7.

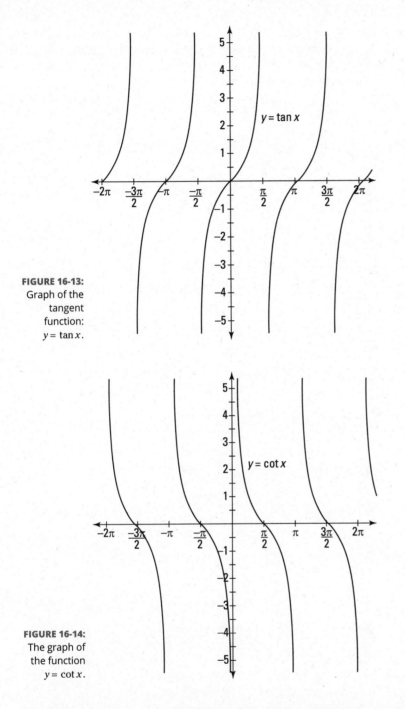

FIGURE 16-13:
Graph of the
tangent
function:
$y = \tan x$.

FIGURE 16-14:
The graph of
the function
$y = \cot x$.

Lining Up and Going Straight with Lines

Straight lines are very invariant; you can perform all the transformations, translations, and reflections you want, but they still remain straight. These lines are graphs of linear functions of the form $y = ax + b$. Figure 16-15 depicts the line $y = x$. In applications, scientists often approximate those curvy (called *nonlinear*) functions using a collection of linear segments that represent linear functions. You'll see plenty of this in a calculus course. The V-shaped absolute function $y = |x|$ is simply two linear functions pieced together. Indeed, all computers draw curves by connecting points on the curves with line segments. So, in some senses, linear functions can be viewed as the parents of all functions.

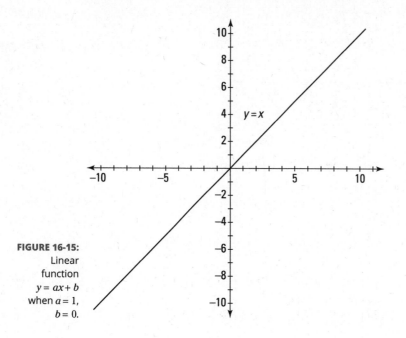

FIGURE 16-15:
Linear
function
$y = ax + b$
when $a = 1$,
$b = 0$.

Chapter **17**

Ten Missteps to Avoid

One the more popular video games from a few years back featured a little guy with a square head running through the square jungle, swinging on square vines, and jumping over square alligators in square swamps. He was avoiding the pitfalls of the jungle. Consider this chapter the vine you can use to jump over the pitfalls that normally trip up the pre-calculus student. And you don't have to be a video game geek to understand this chapter!

Going Out of Order (of Operations)

Operations in an expression or an equation aren't all meant to be done from left to right. For example, $3-7(x-2)$ doesn't equal $-4(x-2)$ or $-4x+8$. You're supposed to do multiplication first, which means distributing the -7 first $3-7(x-2)=3-7x+14$. And then you combine like terms to get $-7x+17$.

REMEMBER

Remember your order of operations (PEMDAS) all the time, every time:

Parentheses (and other grouping devices)

Exponents (and roots)

Multiplication and **D**ivision, from left to right in order as you find them

Addition and **S**ubtraction, also from left to right

To further review the order of operations, see Chapter 1.

FOILing Binomials Incorrectly

REMEMBER

When multiplying binomials, always remember to multiply them in the correct order. You remember FOIL — First, Outside, Inside, Last. This includes when squaring any binomial. The biggest mistake made in these situations is something like when squaring a binomial: $(x-4)^2$ and getting $x^2 + 16$. That's forgetting a whole lot of multiplying, and it's not correct. It should look like this: $(x-4)^2 = x^2 - 4x - 4x + 16 = x^2 - 8x + 16$. You may use other orders, but it pays to be careful and consistent.

Breaking Up Fractions Badly

WARNING

Don't fall for a big trap and break a fraction up incorrectly. $\frac{5}{4x+3}$ doesn't equal $\frac{5}{4x} + \frac{5}{3}$. If you're skeptical, just pick a value for x and plug it into both expressions and see whether you get the same answer twice. You won't, because it doesn't work. On the other hand, $\frac{4x+3}{5} = \frac{4x}{5} + \frac{3}{5}$. See the difference?

When reducing fractions, each term in the fraction has to be divided. The division bar is a grouping symbol, and you have to simplify the numerator and denominator separately before doing the division. For example, $\frac{6}{3x+12} \neq \frac{\cancel{6}^2}{\cancel{3}x+12} = \frac{2}{x+12}$. Instead, you factor the denominator, first, and then reduce: $\frac{6}{3x+12} = \frac{\cancel{6}^2}{\cancel{3}(x+4)} = \frac{2}{x+4}$.

Combining Terms That Can't Be Combined

WARNING

Yet another frequent mistake occurs when terms that aren't meant to be combined are combined. $4x - 1$ suddenly becomes $3x$, which it's not. $4x - 1$ is simplified, meaning that it's an expression that doesn't contain any like terms. $3a^4b^5 + 2a^5b^4$ is also simplified. (It can be factored, but it's still simplified.) Those exponents are close, but close only counts in horseshoes and hand grenades. When counting in the real world, you can't combine apples and bananas. Four apples plus three bananas is still four apples and three bananas. It's the same in algebra: $4a + 3b$ is simplified.

Forgetting to Flip the Fraction

When dealing with complex fractions, remember that you're dealing with division of fractions. $\dfrac{\frac{3}{x+1}}{\frac{2}{x-2}}$ doesn't become $\dfrac{3}{x+1} \cdot \dfrac{2}{x-2}$. Remember that a division bar is division. To divide a fraction, you must multiply by the reciprocal of the denominator, so $\dfrac{\frac{3}{x+1}}{\frac{2}{x-2}} = \dfrac{3}{x+1} \cdot \dfrac{x-2}{2}$.

Losing the Negative (Sign)

It's true that in life you're not supposed to be negative, but in math, don't disregard a negative sign — especially when subtracting polynomials or raising to powers.

$\left(4x^3 - 6x + 3\right) - \left(3x^3 - 2x + 4\right)$ isn't the same thing as $4x^3 - 6x + 3 - 3x^3 - 2x + 4$. If you do it that way, you're not subtracting the whole second polynomial, only its first term. The right way to do it is $4x^3 - 6x + 3 - 3x^3 - (-2x) - 4 = 4x^3 - 6x + 3 - 3x^3 + 2x - 4$, which simplifies to $x^3 - 4x - 1$. The issue here is a special case of the failure to correctly apply the distributive law; it frequently occurs with other coefficients as well (not just –1). Failure to write the parentheses often directly contributes to these errors.

Similarly, when subtracting rational functions, take care of that negative sign.

$$\frac{3x + 5}{x - 2} - \frac{x - 6}{x - 2} \neq \frac{3x + 5 - x - 6}{x - 2}$$

What happened? Someone forgot to subtract the whole second polynomial on the top. Instead, this is the way to do it:

$$\frac{3x + 5}{x - 2} - \frac{x - 6}{x - 2} = \frac{3x + 5 - (x - 6)}{x - 2} = \frac{3x + 5 - x + 6}{x - 2} = \frac{2x + 11}{x - 2}$$

And one other caution about negative signs is when raising negative expressions to powers. There's a difference between -2^6 and $(-2)^6$. In the expression -2^6, the order of operations says to raise the 2 to the sixth power, first and then find the opposite: $-2^6 = -64$. The expression $(-2)^6$ says to raise the number –2 to the sixth power: $(-2)^6 = 64$.

Oversimplifying Roots

When it comes to roots, there are all kinds of errors that can occur. For instance, $\sqrt{4 + 9}$ suddenly becomes 5 when that's not even close! You know that $\sqrt{4 + 9} = \sqrt{13}$.

REMEMBER

Don't add or subtract roots that aren't like terms, either. $\sqrt{7} + \sqrt{3}$ isn't $\sqrt{10}$, now or ever. They're not like terms, so you can't add them. But how about $\sqrt{8} - \sqrt{2}$? You can rewrite it as $\sqrt{4 \cdot 2} - \sqrt{2} = 2\sqrt{2} - \sqrt{2}$ and then subtract the like terms getting $\sqrt{2}$.

Executing Exponent Errors

When multiplying monomials, you don't multiply the exponents.

$x^4 \cdot x^3 = x^7$, not x^{12}. Don't mistake the multiplication of exponential expressions with powers of exponential expressions. Raising $\left(x^4\right)^3 = x^{12}$. Also, when there's more than one term being raised to the power, you have to multiply the whole expression the number of times shown with the exponent. Raising $\left(x^2 - 2x + 3\right)^3$ does not mean $\left(x^2\right)^3 - (2x)^3 + 3^3$. Instead it's $\left(x^2 - 2x + 3\right)^3 = \left(x^2 - 2x + 3\right)\left(x^2 - 2x + 3\right)\left(x^2 - 2x + 3\right)$.

Ignoring Extraneous

Sometimes, when solving radical or rational equations, you have to perform an operation (such as multiplying all the terms by a variable) which can introduce an extraneous or false answer.

For example, when told to solve $\sqrt{4x+1}+x=1$ you dutifully subtract x from each side and then square both sides of the equation: $\left(\sqrt{4x+1}\right)^2=(1-x)^2$ or $4x+1=1-2x+x^2$. Simplifying, you get $0=x^2-6x=x(x-6)$. This new equation has two solutions: $x=0$ and $x=6$. When substituting back into the original equation, the $x=0$ works just fine: $\sqrt{4(0)+1}+0=1$ becomes $\sqrt{0}+1=1$. But that's not the case with $x=6$: $\sqrt{4(6)+1}+6=\sqrt{25}+6=11\neq1$. The solution $x=6$ is extraneous.

Misinterpreting Trig Notation

The trig functions are great to work with, and they come with some specialty notation that makes writing about them quicker and easier. You just need to be careful and interpret the notation correctly.

The two cases in point have to do with squaring the function or finding the inverse of the function.

When you see $\sin^2 x$, this means to find the sine and then square the result.

$$\sin^2 x=(\sin x)^2,\text{ not }\sin x^2$$

$$\sin^{-1}x=\arcsin x,\text{ not }\frac{1}{\sin x}$$

Index

trigonometric identities *(continued)*

  product-to-sum identities, 181–182, 190

  proofs, 165–166, 171–173

  Pythagorean identities, 159–160, 167–168

  reciprocal and ratio identities, 157–158, 167

  sum and difference identities, 175–177, 186–187

  sum-to-product identities, 182–183, 190–191

trigonometric term, 127

Trigonometry For Dummies (Sterling), 3

trinomial polynomials, 76

U

union symbol, 28

unit circle, 121–126

  coordinate plane, 121–122

  defined, 121

  example questions and answers, 122

  finding ratios from angles on, 124–126, 134–135

  point-in-plane, 121

  workbook questions, 122–123

  workbook solutions, 133–134

V

variables, solving for, 10–12

  absolute value equation, 10

  example questions and answers, 11

  isolating absolute value, 10–11, 20–21

  multiple absolute value terms, 20

  workbook questions, 11–12

  workbook solutions, 19–21

vertical asymptotes, 49–51

  example questions and answers, 50–51

  workbook questions, 50–51

  workbook solutions, 65–68

vertical shrinking (flattening), 44–45

vertical stretching (steepening), 44–45

W

Wiley Product Technical Support, 3

X

x-axis symmetry, 40

Y

y-axis symmetry, 40

Z

zero product property, 9

About the Author

Mary Jane Sterling is also the author of *Algebra I For Dummies*, *Algebra II For Dummies*, *Math Word Problems For Dummies*, *Business Math For Dummies*, *Linear Algebra For Dummies*, and *Finite Math For Dummies*. She taught junior high school and high school math for several years before beginning her 35 year tenure at Bradley University in Peoria, Illinois. Mary Jane especially appreciated being able to work with future teachers and trying out new technology. She and her husband, Ted, enjoy spending their leisure time with their children and grandchildren and finding new destinations for their travels.

Dedication

I'd like to dedicate *Pre-Calculus Workbook For Dummies* to my grandchildren: Elliott, Fiona, Wolf, and Blake. They each get copies of this workbook, although it'll be just a few years before they'll be able to or want to read it.

Authors' Acknowledgments

I want to sincerely thank project editor Christopher Morris for all his work and dedication to this endeavor. Thank you to Doug Shaw for keeping my mathematics in check (or checking my math?). And thank you to Lindsay Lefevere for her continued efforts to keep this author creatively and happily occupied.

Publisher's Acknowledgments

Executive Editor: Lindsay Sandman Lefevere

Project Editor: Christopher Morris

Copy Editor: Christopher Morris

Technical Editor: Doug Shaw, Ph.D.

Production Editor: Magesh Elangovan

Cover Image: © Unconventional/Shutterstock